Swansea Copper

SWANSEA COPPER

A GLOBAL HISTORY

Chris Evans and Louise Miskell

Johns Hopkins University Press
Baltimore

Printed in the United States of America on acid-free paper
2 4 6 8 9 7 5 3 1

Johns Hopkins University Press
2715 North Charles Street
Baltimore, Maryland 21218-4363
www.press.jhu.edu

Cataloging-in-Publication Data is available from the Library of Congress.

A catalog record for this book is available from the British Library.

ISBN-13: 978-1-4214-3911-2 (hardcover)
ISBN-13: 978-1-4214-3912-9 (ebook)

CONTENTS

TABLES AND FIGURES

TABLES

FIGURES

ACKNOWLEDGMENTS

We would like to thank the bodies that funded our research over many years. Chris Evans is grateful to the Leverhulme Trust for the International Network Grant that supported the project "A World of Copper: Globalising the Industrial Revolution" in 2012–2013; Riksbankens Jubileumsfond for an award under its Early Modern Modernities program, held jointly with Göran Rydén, University of Uppsala, on "Places for making and places for taking: metals in the global eighteenth century" in 2013–2016; the Center for the History of Business, Technology, and Society at the Hagley Museum & Library for an H. B. du Pont Fellowship in 2011 that enabled him to study the electrical uses of copper in the nineteenth century; and the Faculty of Business and Society, University of South Wales. Louise Miskell is grateful to the Economic and Social Science Research Council for funding a project on "History, heritage and urban regeneration: the local and global worlds of Welsh copper" (2010–2011). Led by Louise's colleague Huw Bowen at Swansea, this initiative was responsible for renewing and widening interest in Swansea Copper in academic, public, and heritage circles. The College of Arts and Humanities at Swansea University has since shown sustained commitment to supporting ongoing research and related public engagement activities.

We have benefited greatly from exchanges with a range of scholars. These include participants in the Leverhulme-funded "World of Copper" workshops masterminded by Olivia Saunders and held at Swansea (facilitated by Huw Bowen), Burra (facilitated by Greg Drew), and Santiago de Chile (facilitated by Luis Ortega). We also wish to acknowledge the stimulus provided by the project "Copper in the Early Modern World: A Comparative Study of Work and Everyday Life in Falun and Røros," led by Kristine Bruland and Göran Rydén, and funded by the Research Council of Norway. We have enjoyed the opportunity to share research findings at Røros in April 2016; at the

National Waterfront Museum, Swansea, in February 2017; at the Fifth European Congress on World and Global History, Budapest, in September 2017; and at the Eighteenth World Economic History Congress in Boston, in August 2018.

Many people have contributed to this ongoing conversation. They are Roberto Araya-Valenzuela, David Bannear, Peter Bell, Peter Birt, Huw Bowen, Kristine Bruland, Roger Burt, Eduardo Cavieres, Des Cowman, Mel Davies, Greg Drew, Lynn Drew, Jay Fell, Johan García Zaldúa, Igor Goicovic, Tehmina Goskar, Steve Hughes, Ragnhild Hutchison, Bill Jones, Gerhard de Kok, Miroslav Lacko, Tim LeCain, Manuel Llorca-Jaña, Denis Morin, Jeremy Mouat, John Morris, Katherine Morrissey, Jürgen Nagel, Juan Navarette-Montalvo, Jonas Monié Nordin, Sven Olofsson, Luis Ortega, Philip Payton, Jorge Pinto, Jennifer Protheroe-Jones, Kristin Ranestad, Inés Roldán de Montaud, Göran Rydén, Sharron Schwartz, Ryuto Shimada, Jason Shute, Keith Smith, Klaus Weber, Alf Zachäus, and Nuala Zahedieh. Our thanks are due to the expert staff at archives and record offices where some of the major collections relating to Swansea Copper are held, especially West Glamorgan Archive Service, the Richard Burton Archives, Bangor University Archives, Bristol Archives, and the National Library of Wales. We would also like to thank Claire Evans of USW Design for the map that appears as figure I.1.

Finally, we are grateful to the editors of the *Welsh History Review* for permission to repurpose material that originally appeared in Chris Evans, "El Cobre: Cuban Ore and the Globalization of Swansea Copper, 1830–1870," *Welsh History Review* 27, no. 1 (2014): 112–131.

Swansea Copper

Swansea in South Wales was to copper what Manchester was to cotton or Detroit to the automobile.

The Swansea District, as it was known in its sulfurous, smoke-choked heyday, was a small part of a small country. It could be walked across, east to west, in less than a day. But what began there at the start of the eighteenth century was big with consequences. It became the place where a revolutionary new method of smelting copper, later to be christened the Welsh Process, began to flourish. Using mineral coal as a source of energy, Swansea's smelters were able to produce copper in volumes that were quite unthinkable in the long-established smelting centers of central Europe and Scandinavia that remained faithful to wood and charcoal. The effects were startling. After some tentative first steps, the Swansea District became a smelting center of European importance. After 1750 it became a center of global significance. Indeed, between the 1770s and the 1840s the Swansea District routinely produced a third of the world's smelted copper, sometimes more.[1]

Swansea's copper industry represented a radical technological departure. The use of mineral coal was at the heart of this change, but new organizational and spatial arrangements were also involved. The smelting of ores, the great French metallurgist Frédéric Le Play observed in 1848, had always taken place in close proximity to the mines from which those ores were raised—a universal law that had been respected since prehistory. Because metallic ores were so dense, fuel was carried to the ore. But the costs of transporting fuel, whether firewood or charcoal, were themselves considerable. As a result, traditional mining-smelting centers drew upon forest resources that were immediately circumjacent, making smelting a localized process. The Swansea model was revolutionary because ore was brought to the fuel. Indeed, ore was carried a considerable distance. The major source of copper ore in eighteenth-century Britain was Cornwall. Because Cornish mines were located within striking distance of the coast, ore could be shipped across the Bristol Channel to Swansea Bay (figure I.1). Seaborne ore was a distinguishing feature of the Swansea model. It made Swansea—the smelting center without any ore of its own to smelt—an industrial town that broke with every precedent.

Swansea and Cornwall were codependent. Cornish mines supplied the ore for Swansea's coal-fired furnaces and the returning ore barks were loaded with Welsh coal to power the steam engines that drained Cornwall's mines. Cornish ores were supplemented by imports from Ireland and very occasional shipments from colonial America (which appear to have been no more than experimental), but Cornwall's supremacy went unchallenged until the 1770s, when a hitherto unexploited ore body on the island of Anglesey, off

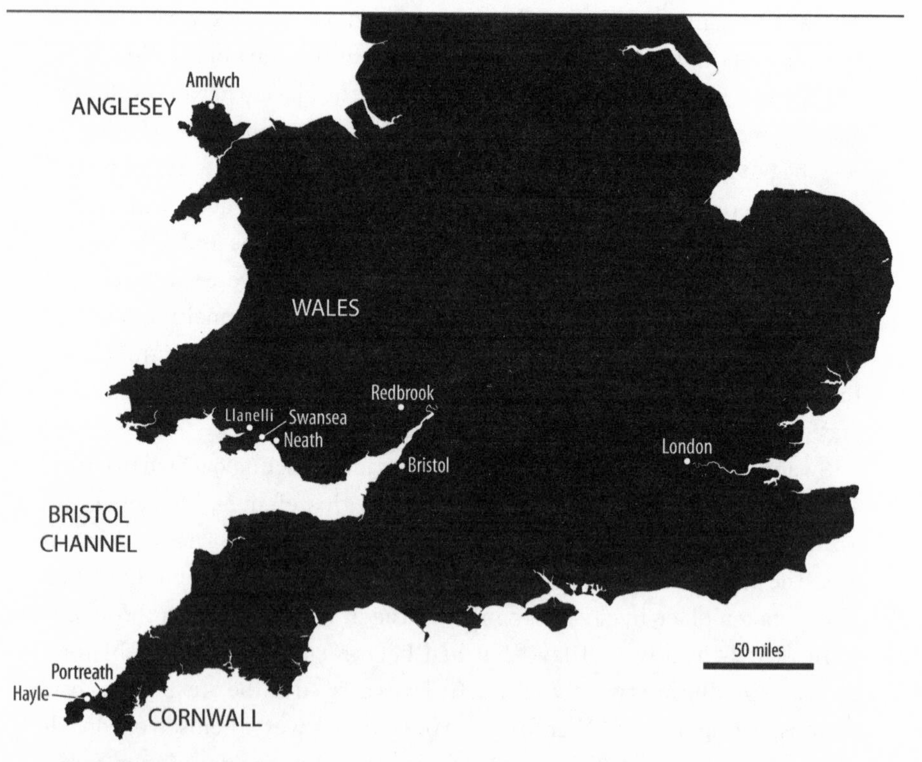

Figure I.1. Location of the Swansea District and related centers of the copper trade in England and Wales
Claire Evans, USW Design

the north coast of Wales, came onstream. The ores of Parys Mountain could be extracted by open-cut methods, giving them a distinct advantage over the deep-mined products of the English Southwest. The Cornish mining sector was thrown into crisis. Had the Anglesey deposits not been exhausted within a generation the crisis might have been more pronounced. As it was, Cornwall resumed its dominance in the first years of the nineteenth century. It was not challenged again until the 1830s.

In theory, copper ore could be shipped in to Swansea from any point on earth provided it was of sufficient richness to bear the costs of freight, but tariff barriers stood in the way. However, a relaxation of British customs regulations at the end of the 1820s brought about a striking diversification of the supply chain. Ore barks that had hitherto been restricted to British and Irish waters now left for distant harbors. All of a sudden, Frédéric Le Play remarked, the Welsh copper sector seemed to know "no limits other than those of the globe itself." Swansea received ores "from the island of Cuba, from Mexico, from Colombia, from Peru, from Chile, from Australia and from New Zealand."[2] With this development, early Victorian Swansea became far more than an industrial center of note; it became "Copperopolis," the hub of a global production network, mobilizing capital, labor, and technology over immense distances. Cornish miners were dispatched to Cuba, railway engineers headed to Chile, and smelting specialists sailed for Adelaide, South Australia.

"Swansea Copper" (with a capital *C*) is our shorthand for this mode of metal production. Swansea copper (uncapitalized) was the metal that originated in the Swansea District. It is a term we use infrequently. Swansea Copper denotes the tangle of furnaces and ore yards that spread across the Swansea District in the eighteenth and nineteenth centuries. Swansea Copper was an industrial organism, powered by coal, given life by hundreds of copper workers and colliers, and set in motion by metropolitan capital. By extension, Swansea Copper took in both the Swansea District and the mining zones to which it was joined, initially in Cornwall, then Anglesey, and then, after 1830, much farther afield. Swansea Copper therefore also includes the maritime links that connected the Swansea District to its increasingly distant mineral tributaries: the fleet of copper barks and those who crewed them.

As an articulated production network of transcontinental reach, Swansea Copper helped globalize Britain's Industrial Revolution between the 1830s and 1850s. But the very success of Swansea helped undermine its hegemonic position. Refinements to the Welsh Process began to make it less profligate in its use of coal. But with every advance in fuel efficiency, the Welsh Process became progressively detached from coal-rich South Wales.[3] The spread of the railroad also contributed to a shaking up of the geography of nineteenth-century copper production. Hitherto, smelting had been conducted where fuel and ore were in close proximity, or, in the exceptional case of Swansea, where furnace stuff could be transported by sea to a coal-rich location. The railroad offered more varied options. Minerals could now be shifted to and fro in more flexible patterns. The first smelters on America's eastern seaboard, for example, brought in their ore by bark, Swansea-fashion, but their coal came via rail. A Swansea-centric world could now give way to something that was far more polycentric. In the 1860s these new tendencies came powerfully to the fore. The opening up of new ore fields in the American interior marked a decisive change. Technologically, the new American copper industry drew upon earlier Welsh practice; the first calciners and reverberatory furnaces to be built in Montana adhered to the Welsh pattern. But they soon assumed a more gigantic form, and before the nineteenth century was out the American copper industry had adopted smelting and refining methods that departed from the Swansea tradition.

Thus, in the 1860s Swansea's period of global dominance came to a close. The Swansea District remained an important smelting center for decades to come. Indeed, its output of smelted copper did not peak until the 1890s. But Swansea's share of global production shriveled rapidly in the final third of the nineteenth century. The new production facilities in the Rocky Mountain West saw to that. Increasingly, Swansea's role was to be that of a specialized refining and processing center. The smelting of ore finally ended in the 1920s. By the late twentieth century some weathered and overgrown ruins and acres of slag still too toxic for most plant species to tolerate were all that remained.

~

What can this long developmental arc tell us? The Swansea District may have been small, but it speaks to a more expansive history, a history that spans oceans and embraces different continents. Indeed, the history of Swansea Copper, if it is to be anything, must be a contribution to global history. Copper has much to offer global historians, scholars who are habitually concerned with connections across wide areas of space. Metals, which tend to have high value and are not generally perishable, are highly connective. They have been traded over long distances for millennia. The very fact of a Bronze

Age presupposes that copper and tin were circulating in metallic form in the fourth millennium BCE. The production of bronze as an alloy required that its constituent metals were melted together.

Metals traveled; ores (the rock from which metal was extracted) seldom did. It would be wrong to say that ores *never* traveled. Brilliantly colored minerals like malachite or azurite (both forms of copper carbonate) have circulated from prehistory to more recent times as precious objects to be put to decorative or ritual use. But circulation in this instance was governed by their crystalline rarity and portability. Ores that were to be smelted, on the other hand, and that were necessarily required in bulk, were seldom transported any great distance. Smelting, as Le Play observed, was an activity tied to ore bodies. Metals, by contrast, were necessarily mobile because there is no reason why mineral lodes should coincide with human settlement. Humans congregate where game is abundant or soils are rich. Metals have always, therefore, had to be taken from where they were found to where they were wanted. This was true even of iron, whose ores are distributed widely; it has been doubly true of metals like tin or copper, which are very much scarcer. Geology has always been a hard master, as scores of medieval and early modern prospectors and assayers could testify. Copper might have to be mined at high altitude (as was common in Alpine Europe) or at latitudes that made year-round exploitation impossible (as was the case with Kengis in Sweden, opened in 1644 within the Arctic Circle).[4]

By the early modern era copper was traded globally. Copper was known in pre-Colombian America, albeit on a very limited basis. "Native copper"—copper that occurs naturally in its pure metallic form—was to be found in the Great Lakes region.[5] Its luster and mysterious malleability gave it a sacred quality. As such, it featured in trade and diplomatic exchanges up and down the Mississippi Valley and along the St. Lawrence. Not surprisingly, native peoples welcomed the copper vessels that Europeans brought with them, valuing them as much for their spiritual power as their utility. Copper had potency in the Old World too. "Red gold" was held in high cultural regard in sub-Saharan Africa, where it had been smelted since the first millennium BCE.[6] When Portuguese traders arrived on the Guinea coast in the fifteenth century they soon discovered that copper was a very acceptable article of trade with African merchants. Copper was to remain a significant element in Afro-European exchange for centuries thereafter.[7] In Asia, Japan was a major exporter of copper, shipping bars from Nagasaki to China and the Indian Ocean world. By the seventeenth century South Asia was also being supplied with copper from Europe as well as Japan, mostly via the Dutch East India Company, although if prices spiked in Europe the flow of materials might be reversed and it would be Japanese copper that was shipped to Amsterdam.[8]

The mobility of metals suggests a need to rewrite our narratives of emergent modernity. The new global interactions of the sixteenth century that interlinked the Indian, Atlantic, and Pacific oceans are commonly understood as being powered by the trading of spices and precious fabrics. Metals do not feature in this story, unless it is the silver of Potosí, shipped east via Spain or west via Manila to pay for the cinnamon and silks. Our understanding of global change in the seventeenth and eighteenth centuries follows suit. The most striking historiographical advances of recent years have come through the study of globally traded goods that were intended to add to domestic comfort, to enhance social prestige, or to figure in acts of exotic consumption: Indian cottons, Chinese ceramics, or foodstuffs like sugar or tea. Historians have been interested in how these goods led to novel forms of social interactions and a new sense of the self. Distinctively modern forms of social personality, it is said, were generated in acts of consumption, because a broadening and diversifying of consumption practices proved incompatible with conceptions of the social order that looked to immutable hierarchies of rank or order. Consumer goods, in short, were liberating, thus justifying their study. It has been hard to make the same claim for producer goods. Whereas handsome items of furnishing lend themselves to psychocultural interrogation, a stack of sawed timber does not; nor does barreled tar, coiled hemp, or coked coal. Nor, for that matter, do ingots of metal. Metals register in the new historiography of consumer taste if they come in an especially bijou form—as ormolu gilding, for example, or as pinchbeck trinketry—but metals in their raw form have been crowded out of the picture. Yet producer goods need to be brought into the picture, not excluded. The proliferation of new consumer goods can only be understood in relation to the producer goods from which they were fashioned, and those producer goods were becoming available in greater volume and in improved and more diverse forms. The success of the small metalware trades in Georgian Britain rested upon commodity chains that stretched east to the Ural frontier of the Russian Empire and west to the Chesapeake. A productive web that extended halfway around the northern hemisphere allowed artisans in London, Birmingham, and Sheffield to draw upon highly specialized feedstock that included German steel, Russian bar iron, and colonial pig iron.[9] The sort of metallic items that might be found in a polite drawing room (a tea urn, for example, or sugar nippers) could trace their origins to remote forests or distant mining camps. Historians have rarely traced those connections.

Writing producer goods such as copper back into our narrative of modernity is difficult because producer goods typically arise from processes that are large in scale, often dirty, and sometimes toxic. They issue from mines, furnaces, sawmills, and kilns. They are redolent, in other words, of the Industrial Revolution in its gritty, clanking, classical form. Yet the traditional conception of the Industrial Revolution, which stressed its rapidity, its reliance upon revolutionary technological innovations, and its essential Britishness, has fewer and fewer champions. Far from being the harbinger of modernity, British industrialization is now painted as an odd, thoroughly atypical phenomenon. Historians are loath to focus on the British Isles between 1760 and 1830, the bookend dates of T. S. Ashton's classic account of the Industrial Revolution; they look instead to developments that extend across a broader Eurasian/Atlantic space and that stretch from the sixteenth century to the nineteenth.[10] Historians have also stepped away from the technological explanations of change that held sway in the mid-twentieth century (in the popular mind at least). Supply-side explanations of the coming of modernity focused on the gadgets and machines that still stand as mileposts in textbook histories of the Industrial Revolution—the flying shuttle, for example, the spinning jenny, the Newcomen engine, or the puddling furnace. In the last decades of the twentieth century this concentration on clattering machines and smoking chimneys fell from fashion, and demand-side explanations came into vogue. Careful research on household inventories revealed a broadening range of furnishings and utensils within prosperous households in England well ahead of the conventionally understood Industrial Revolution. Other historians investigated retailing and marketing in Georgian England and uncovered innovations in the distribution and display of consumer articles. The branding and advertising of goods was not, it turned out, a phenomenon of modern times; it belonged to the eighteenth century. The striking feature of the age, it began to be argued, was not process innovation (the invention of new ways of manufacturing) but product innovation (the development of new things).

Given this shift in thinking, it might seem as though Swansea Copper has been historiographically orphaned. The Industrial Revolution is routinely dismissed as a defunct historical concept, while global history, which emerged as a lively conceptual field just as the Industrial Revolution fell fallow, has little time for producer goods. Indeed, global history would at first sight have little to offer students of Swansea Copper. The global turn in historical scholarship became pronounced in the 1990s, spurred on by the collapse of Soviet communism, the rise of the East Asian economies, and the onset of the digital revolution. Indeed, information technology seemed ready to enable (if not enforce) a common global experience. The internet, in particular, emerged as a technology that could erase the boundary between the local and global; it opened the way to a borderless digital sphere. Meanwhile, the East Asian boom reminded scholars that global commerce had once pivoted about China; in view of the growth rates that were being achieved in the People's Republic at the start of the twenty-first century, it seemed set to resume its ancient track. East Asia was a strong corrective to Eurocentric narratives of the past. Taken together, these developments struck hard at the presumption that the European, much less British, experience was normative. They also cast doubt on narratives of the past that celebrated production. Soviet communism had lauded work as an ennobling collective experience, but post-1989 this was revealed as work without reward. Communism had given rise to lumbering state systems that had manifestly failed to satisfy the consumer needs of their citizens. The ending of the Cold War seemed to install consumer capitalism as humanity's destiny, or at least its default mode. The opening of the digital age seemed to promise consumerism of a peculiarly weightless kind—a riot of visual and auditory stimuli, of disembodied experiences. The parochial or vernacular was easily detached from its local moorings and employed in customized or hybridized ways by users who had no knowledge of the original vernacular setting. Social scientists hailed this as "globalization," a term that was installed as the key discursive term of the 1990s.

Historians responded to these heady developments by examining premonitions of modern globalization. They identified earlier phases of global interaction—"archaic globalization" or "proto-globalization."[11] They concerned themselves with the circulation of people and things in earlier times. This was all of a piece with the broader historiographical trends that prioritized consumption and looked to the acquisition of high-value articles as a motor of social change. Global history might then appear to make the marginalization of Swansea Copper, an irredeemably sooty phenomenon, complete. Yet appearances can be deceiving. In fact, the practice of global history offers a way of reconceptualizing one of the quintessential British industrial experiences. Although global history initially gave a further fillip to demand-side explanations of change, it has necessarily raised questions about production processes and how best to reconcile supply and demand, production and consumption, in writing our narratives of emergent modernity.

After all, the busy long-range trade in manufactured items that is now recognized as a key feature of the early modern period presupposed the existence of specialized manufacturing zones across Eurasia, Africa, and the Americas. Freed from the assumption that industrial dynamism was a European preserve, global historians are now conscious of great industrial cities like Jingdezhen, the ceramics capital of China, whose kilns, which numbered in the hundreds, honeycombed urban hillsides. They are attentive to previously underappreciated landscapes of textile manufacturing as well: the networked spinners and weavers who could be found in Bengal making fine muslins or across the Sahel of West Africa weaving *pano* cottons. The history of cotton textiles has, in fact, played a critical role in the development of global history as a practice. Indian cottons enjoyed an enormous vogue in seventeenth- and eighteenth-century Europe. They were produced with a delicacy that European manufacturers could not match, and colored in ways that were ravishing to the European eye. In that sense, cotton textiles lent themselves to analyses that saw them as items of fashion, using a consumerist lens. But cottons can also be used to account for changes in production patterns. South Asian textiles were so lucrative that they spawned imitators. European manufacturers sought ways of counterfeiting muslins from Bengal or "chints" from the Malabar Coast.[12] They were encouraged to do so by governments that were concerned that the rage for Indian cottons came at too a high cost, through the drain of bullion to the east. Indian goods were highly desirable but they appeared to jeopardize national finances. Import substitution—the making of ersatz "Indian" textiles in Europe—was the answer. English cotton goods might not have matched the quality of Indian goods, but English willingness to mechanize production processes made them cheaper. Labor-saving devices sent productivity vaulting upward, none more than the spinning frame of Richard Arkwright: "wee shall not want 1/5 of the hands I First Expected," Arkwright exclaimed as the full potential of his machinery dawned upon him.[13] The effects were world changing. At the start of the eighteenth century India was the world's leading producer of cotton textiles. By the century's end that position of dominance belonged to Britain. "The slow Progress of an Indian Manufacture," a committee of the East India Company reported in the 1790s, "unaided by Machinery, will require Ten, Twelve, perhaps Fifteen Persons to perform the same Work which a single British Manufacturer can execute, assisted as he is by numerous Inventions and Improvements."[14]

Seen in this way, European industrialization had its roots in the East. The rise of cotton manufacturing in Lancashire was catalyzed by the older, more sophisticated cotton manufactures of South Asia, just as the porcelain of Sèvres or Meissen was imitative of superior Chinese ware rather than a testament to European expertise. Swansea Copper might usefully be seen in this light too. Swansea smelters were aware of Japanese copper's historic dominance over Asian markets, and when Welsh copper began to be shipped eastward by the East India Company they sought to mimic the vivid scarlet of Japanese bars. Michael Faraday, no less, who visited Swansea in 1819, witnessed copper being prepared for the Chinese market. The molten metal was poured into molds and then immediately turned out into a tank of cold water. "Cast in this way, the ingots assume a very brilliant red colour and are sold to the Chinese as Japan copper."[15]

Indeed, the fortunes of Swansea Copper were from the outset closely linked to global markets. The emergence of a British copper industry in the late seventeenth century coincided with a major expansion of the Caribbean sugar sector, which provided a dynamic new market for copper vessels, whether sugar pans or stills.[16] Starting in the 1730s India absorbed a growing quantity of Swansea Copper (often in the guise of "Japanese" copper). Then, in the 1770s, came a further dramatic new pulse of demand in the form of sheathing for naval and merchant vessels that plied warm waters.[17] Repeatedly, Swansea's success stemmed from new sources of demand that were colonial or military in character. Buoyant domestic demand was helpful, of course, but external demand bulked up orders and—critically—affected technological development. A Swedish observer who watched "Guinea rods" being made for the African market in the 1750s thought the technique adopted was wrongheaded. A more expeditious and economical method was available, but that method would not give the metal the ductility that Africans prized nor the surface patterning that they found pleasing.[18] The globalization of Swansea Copper in the 1830s and 1840s, when copper barks began to bring in ores from the southern hemisphere, was an astonishing extension of Swansea's supply chain, but Welsh smelters had always had horizons that were global.

Swansea's hegemony was based upon coal. Indeed, Swansea Copper was prodigal of coal, blithely so, it seemed, at a time when other smelting centers husbanded their energy supplies with meticulous care. This reliance on coal has an importance for current views of global history. One of the most influential recent interventions in the debate over the origins of (western) modernity has come from a historian of China, Kenneth Pomeranz, who posed the question of why China, so long the epicenter of the global economy, was eclipsed by the West in the nineteenth century. In 1750, so Pomeranz maintained, there was little to choose between the more developed zones of China—the Yangtze River delta, say—and the most advanced regions of Northwest Europe. Both were highly entrepreneurial societies, with efficient market mechanisms and security of property. Both were technologically creative and both hosted populations that enjoyed high living standards. In the eighteenth century, Pomeranz suggested, there was little to separate the advanced societies that flourished at the western edge of the Eurasian landmass from those that flourished in the East. That industrial modernity should burst forth on the western extremity of Eurasia was not inevitable. Yet in the nineteenth century that is what happened. The Great Divergence, as Pomeranz christened it, saw Northwest Europe experience giddying growth while China regressed.[19]

What could account for this historic parting of the ways? Nineteenth-century China, Pomeranz explained, succumbed to environmental crisis as population growth led to soil depletion and an overexploitation of forest resources. Political disintegration ensued. How did Europe, which underwent a startling demographic expansion of its own, avoid this fate? Pomeranz offered two answers. First, New World colonies provided a huge acreage of virgin soil which, by offering an external supply of timber, carbohydrates, and animal protein, helped Europe's population to grow without short-circuiting economic expansion. Second, and more importantly in the present context, the easily accessible coal reserves of Northwest Europe allowed for energy to be used lavishly in a way that could not be replicated in wood-dependent Asia. This happy conjuncture enabled Europe to escape economic and ecological involution of the sort that afflicted China.[20]

The relevance of the Great Divergence to the history of Swansea Copper should be clear. Few industrial sectors had such an appetite for energy—eighteen tons of coal were required to produce one ton of copper under the Welsh Process, according to Matthew Boulton of Birmingham—and none switched as swiftly or as emphatically to coal as the British copper industry.[21] In that sense, Swansea Copper exemplifies the Great Divergence. Every other copper-smelting district in the world obtained the energy it needed from vegetable matter. An expansion of smelting required more wood, which implied a greater acreage of timber. But because the surface of the earth is finite, devoting more land to timber meant less land was available for arable purposes or for pasturing animals, and less land that could be used for industrial crops like flax or cotton. A greater supply of copper, in other words, could only come at the expense of a reduced food supply or a shortfall in industrial raw materials. Swansea Copper resolved that ecological conundrum. By tapping subterranean energy Swansea Copper made next to no demands on the surface of the earth. Production could soar upward. The only impediment now was a shortfall in the supply of ore, not a shortage of fuel.

At first sight, Swansea Copper embodies the Great Divergence in a very straightforward fashion, but it also affords us the opportunity of thinking about energy usage and its relationship to modernity in more nuanced ways. For one thing, it can help us break the new global history out of its Eurasian corral. The key works in recent global history tend to focus on the seesaw relationship between the east and west of Eurasia over the last half millennium. The world beyond Eurasia is rather neglected. Swansea Copper, by contrast, is authentically global. It numbered Japanese copper among its competitors in its formative years, it is true, but Swansea Copper always had a frame of geographical reference that extended far beyond Eurasia. It had to contend with "Barbary" copper from North Africa, which was a significant player on the British market in the early eighteenth century. From the start, Swansea Copper reached out to the Caribbean, a key market for as long as the sugar sector flourished there, and it furnished goods for the Guinea trade. Then there was India, into which copper from Swansea poured from the mid-eighteenth century onward. In short, Swansea Copper interacted with the known world, or the world known to commerce, in its entirety. In the nineteenth century, with the world economy ballooning outward and upward, Swansea Copper expanded its range still further. It stretched out to Latin America just as soon as the collapse of Spanish colonial power made direct communication possible: Mexico, Gran Colombia, and, most important of all, Chile. From the 1820s Swansea ore barks battered their way around Cape Horn to Valparaíso, making it as familiar to Swansea's nineteenth-century seafarers as Hayle in Cornwall had been to their forebears a century earlier. In the 1840s the Antipodes became the latest new frontier for Swansea Copper as mineral strikes were made in South Australia and New Zealand. In the 1850s Namaqualand on the northern edge of the Cape Colony became the latest tributary of Swansea District. This was Swansea Copper at its imperial zenith; it spanned the globe.

Coal made all of this possible. It made the reverberatory furnace, the key feature of the Welsh Process, the globally dominant means of copper smelting in the nineteenth century. Swansea Copper's influence was felt everywhere, but that did not mean a slavish replication of the Swansea model. Chile's mines were seldom drained by steam engines in the Cornish style. They relied upon human muscle power: relays of laborers who, with sloshing leather pouches on their backs, clambered hundreds of feet up crude ladders to empty water on the surface. Steam, a massively expensive and uncertain capital investment, could be dispensed with when a surfeit of famished porters were bidding for work. So, in the right circumstances, could coal. The first Welsh-style furnaces to be lit in Australia were fed eucalyptus logs. In the absence of coal the next best energy source would do. In fact, once the Welsh Process had been lifted clear of its Welsh context it proved surprisingly adaptable. Swansea Copper, then, provides an object lesson in how coal technologies might be made to work in new and at first sight unpromising environments. Its history is not one of once-and-for-all triumph; it is a history of modulation and adaptation.

Equally, it has to be said that the technological radicalism of Swansea did not lead to the obliteration of rival production centers. Many of Europe's long-established smelting districts continued to flourish by serving traditional markets. Røros in Norway, established in the 1640s, was still producing high-quality copper in the 1740s and, indeed, in the 1840s. Upper Hungary (modern-day Slovakia) continued to supply the Habsburg lands with copper in the eighteenth century as it had in the seventeenth. Via Trieste, Hungarian copper circulated throughout the central and western Mediterranean, from Venice to Alicante.[22] The eastern Mediterranean was served by the mines of Anatolia: Tokat and Diyarbakir. Indeed, Diyarbakir (southeast Turkey) faced both west and east. Floated down the Tigris to Basra, its copper entered the Indian Ocean world to circulate alongside copper from Swansea.[23] The signal feature of Swansea Copper was not so much its ability to invade traditional markets as its capacity to fulfill new purposes. That was the truly revolutionary road taken at the end of the seventeenth century. Of course, Swansea Copper was formed into pots and pans, and beaten into shapes that would have been as familiar in 1900 as they were in 1600, but it also drove forward the modern, industrial consumption of copper. It made possible the production of stills and sugar boilers on a mass scale, without which the slave-sugar complex of the Caribbean would have been hobbled. It sheathed the hulls of naval and merchant vessels in an age in which the

British established military and commercial supremacy over the world's oceans. Then, in a time of headlong industrial growth, Swansea Copper was embodied in the tubing of steam locomotives, in telegraphic cabling, in the roller-printing of cotton, and in the generation and transmission of electrical power. The course of modern life, it has been said, is "completely dependent on the mechanical, electrical, and magnetic properties of metals."[24] So it is. It is a measure of Swansea Copper's historical centrality.

And yet Swansea Copper's story is so little known. In part that is because its corporate history is so complex. The sites upon which Swansea Copper was smelted were comparatively few, but they were occupied by a bewildering succession of partnerships, some of them, especially in the eighteenth century, lasting for only a few short years. Just working out a sequence of occupation is a daunting, long-term labor. That may explain the empirical character of a good deal of scholarship on Swansea Copper. Some of the key publications display a commanding knowledge of Swansea Copper's kaleidoscopic partnerships but are reticent about construing that knowledge within a broader theoretical literature.[25] For a study of Swansea Copper within a national framework it is necessary to consult Henry Hamilton's *The British Brass and Copper Industry to 1800* (1926).[26] A pathbreaking work in its day, it remains a store of valuable material. But its defects are plain. It is venerable and necessarily fails to speak to present-day historiographical concerns. It closes somewhat arbitrarily in 1800, when Swansea Copper's history was but half done. Moreover, its central thesis, which ascribed the failure of British copper in the seventeenth century to the effects of chartered monopolies and its success in the eighteenth to the striking down of monopoly, has been convincingly rebutted.[27] Nearly a century after the publication of Hamilton's *British Brass and Copper* there is nothing that captures the full span of Swansea Copper or that responds to current historiographical demands. Fine scholarship is available but not always accessible. Some of it is locked away in unpublished doctoral theses, or postgraduate research that has only been partially published.[28]

With *Swansea Copper: A Global History* we attempt to remedy those deficiencies. This book aims to be sweeping in its coverage, ranging from the seventeenth to the early twentieth century. It seeks to establish a geography for Swansea Copper that gives the Swansea District its due but does not neglect the other parts of the commodity chain in which the Lower Swansea Valley was but a single link. It gives proper consideration to mines in Cornwall, Wicklow, and Anglesey, to the battery works ringing Bristol, and to the rolling mills that drew on rivers to the west and south of London. The book attends to the consumption as well as the production of copper. What happened to the sheets of copper that issued from those rolling mills? Were they destined for a distillery in colonial New England, or a sugar boiling house in the Caribbean? Or were they meant for something as familiar and homely as a kettle? Work that explores the distribution, much less the marketing, of copper in Swansea Copper's heyday remains very much the exception. This book also gives systematic attention to Swansea Copper's international reach—an under-researched area.[29] A good deal has been published on Swansea's maritime links, but much of it, especially that devoted to the "Cape Horners" who braved the southern oceans, succumbs to a romanticizing view of its subject. Work that explores the global entanglement of Swansea Copper in a sustained and serious way is still all too rare. The fact that Swansea barks made their way to this or that harbor is celebrated; the impact that mariners or migrant workers had on, say, Chile's Norte Chico, does not command the same level of attention.[30] Above all, this book seeks to open a dialogue between the history of heavy industry and current global scholarship.[31] The history of metals, we contend, can be more than an addition to the corpus of global history; it can offer a different sense of space and time.

~

Our approach in the chapters that follow is broadly chronological. Chapter 1 will examine the place of copper in baroque Europe. It surveys the productive landscape into which Swansea Copper erupted and details the means by which it did so. That will require an account of the Welsh Process, the technological foundation of Swansea Copper. Chapter 2 will focus on the years in which Swansea Copper became firmly established as an important industrial sector in Britain. The links with Cornish mining will be explored, and the sources of capital upon which the first smelting ventures relied will be traced. The origins of the labor force in these formative years will be broached, and the markets to which Swansea's copper was dispatched will be mapped out.

Swansea Copper achieved global dominance in the years between the Seven Years' War and Waterloo. Wales was already a significant producer of copper in 1750, but by 1815 Welsh smelters were responsible for a commanding share of global output. Chapter 3 will account for this upswing. It will take into consideration the impact of new ore bodies, notably those of Anglesey, and assess the importance of new uses for copper, among which the "copper-bottoming" of oceangoing vessels was conspicuous. The following chapter (chapter 4) will be devoted to Swansea Copper in the early Victorian age, when new commercial regulations allowed the importation into Britain of copper ore from around the world. Suddenly, Swansea acquired sources of ore supply that were very distant, initially in Latin America. Chapter 4 will illustrate this development with a study of one of the most important manifestations of Swansea Copper as a global force: the mining settlement of El Cobre in Cuba, where a mixed workforce of enslaved Africans and indentured Chinese laborers was put into the service of Swansea's smelting companies. Chapter 4 will also feature an analysis of labor in the early Victorian copper industry; it will conclude with an account of the crisis in social relations that gripped the Swansea District in the early 1840s and the great strike of 1843 that ensued. The defeat of the furnacemen's strike in 1843 marked a caesura, the consequences of which will be examined in chapter 5. The employers' victory proved pyrrhic. In the mid-1840s, new centers of copper smelting, usually modeled on Swansea and employing refugees from the Swansea District, began to spring up in the Americas and Australia.

The sixth chapter opens in the aftermath of the American Civil War. The war had seen a rapid expansion of the US copper industry, with the development of new ore fields in the American interior. Wartime boom was followed by postwar contraction, leading to convulsions in the international market and precipitating the collapse of some of the landmark features of Swansea Copper, the mining districts of Cornwall and El Cobre among them. Swansea remained an important site for copper production in the last third of the nineteenth century but it was no longer preeminent. Indeed, Swansea Copper was now dwarfed by what was going on in the American West. Whereas Swansea Copper had introduced a seaborne model of smelting, with ores being shipped to southwest Wales, American enterprise reinstated a land-bound model of copper production. Smelting ore in relative proximity to the mines from which it was extracted became the norm once more, first in the Rocky Mountain West and then overseas in Chile and Japan. The spatial rearrangement of copper production in the late nineteenth century was accompanied by the take-up of new technologies that were pneumatic or electrical in character, and that departed from the coal-heavy Welsh Process. A phase of industrial development that had begun two hundred years earlier thereby came to a close. Copper remained a feature of the regional economy in South Wales in the early twentieth century, but the world of Swansea Copper had dissolved.

Copper in Baroque Europe

Copper was the first metal that humankind learned to smelt, perhaps as early as seven thousand years ago.[1] It has been valued for its beauty and utility ever since. It is easily worked and conducts heat readily. For that reason, copper vessels have always been preferred for domestic or industrial processes that require the close control of heat: cooking, for example, or the refining of sugar. Copper's malleability meant that braziers and coppersmiths could form it into an endless array of shapes, both functional and expressive.[2] Indeed, so malleable was the red metal that it could be beaten or rolled into thin tiles and sheets, thereby providing a watertight roofing material. Copper is not an abundant element, however, so its use as cladding has always been restricted to the more prestigious architectural projects: churches, palaces, or the grander sort of civic building. Indeed, because copper is relatively rare (iron is nine hundred times more abundant in the earth's crust) it has been routinely used in past times as specie, as a recognized bearer of value in its own right. Taken together, these qualities ensured that copper and cupreous articles were traded widely across the early modern world.

Copper does have limitations; it is not suitable for foundry work. Viscous when molten, copper "casts poorly, creating objects with disfiguring surface craters caused by offgasing during cooling."[3] When copper is combined with zinc or tin, however, those difficulties abate. Brass, an alloy of copper and zinc, is eminently fusible. So is bronze, the alloy of copper and tin. Brass retains some of the yielding quality of copper, which allows it to be shaped with stamps and dies, and it has something of the luster that has made copper so prized as a decorative substance. Brass, however, does not stand up well to abrasive wear. That is bronze's great strength. It can be cast into forms that will stand recurrent stress, hence its use in bell founding since antiquity and, from the fourteenth century onward, in the casting of cannon.[4]

Copper was then a vital part of the early modern world. Some of the growth industries of the age were absolutely dependent upon it. How would distillers have fared without their cucurbits and alembics of beaten and riveted copper? The manufacturers of armaments were equally reliant on copper. Iron ordnance was cheaper but bronze cannon were lighter and more maneuverable. There were, moreover, few aspects of domestic comfort to which copper and its alloys did not contribute. Brass candlesticks and sconces illuminated this world; andirons and brass fenders gave shape to its hearths; and warming pans took the chill off bedsheets. Copper pans featured heavily in prosperous kitchens, of course, but copper and brass were also formed into the chafing dishes and trivets that kept food warm at the table. Copper coffeepots and teakettles helped keep their owners alert and sociable; cuprous tankards made them companionable and drowsy. There were not many areas of personal life into which copper, one way or another, did not intrude. Professional life was also garnished with copper and brass. They were embodied in the banker's scales, in the bowls in which barber-surgeons collected the blood of their patients, in the locks that secured the merchant's ledgers, and in the quadrants and theodolites put to use by surveyors.

Early Centers of European Copper

There was a rising demand for copper in sixteenth-century Europe occasioned by a growth in the scale of warfare, an extension of navigation, closer links to non-European markets (especially those of West Africa), surging demographic growth, and an elaboration in the material culture of Europe's towns and more prosperous rural districts. The demand was met from mining districts in central Europe: the Tyrol, Thuringia in central Germany, and Upper Hungary, a territory fought over by the Habsburgs and the Ottomans in the early modern period. These regions were caught up in a generalized mining boom that began in the fifteenth century. It is a world still visible to us in the woodcuts of Georgius Agricola's *De Re Metallica* (1556), with its stylized rocky landscapes, cavernous underground workings, and the meticulously depicted wooden machinery that drained mines and pounded ore. Smelted copper from these districts circulated widely. Alpine mining fed manufacturing centers in northern Italy and southern Germany. The Thuringian smelters were strategically well placed to supply customers in southern Germany and the great brass-making centers of Aachen and Stolberg that lay west of the Rhine, while the Elbe provided a highway that took Thuringian copper north to Hamburg and thence to markets in western Europe. Hungarian copper also moved northward along the Elbe. It traveled down the Vistula too. Transhipped at Danzig, it was conveyed onward to Amsterdam or Antwerp.[5]

Central European copper production reached its apogee in the 1560s. In the decades that followed, output and investment tailed off. Because central European copper ores were frequently silver bearing, smelters had prospered through their use of the *Saigerprozess*, a refining process that separated out the copper and silver. Copper boomed between the 1460s and the 1560s because its production was cross-subsidized by sales of silver, but the arrival of large volumes of South American bullion in mid-sixteenth-century Europe depressed silver prices and lowered the return on argentiferous ores. The profitability of copper sank with it, leading to a contraction of mining and smelting across most of central Europe.[6]

Compensation for this shortfall came in the form of Swedish copper, which was to dominate north European markets from the 1570s to the late seventeenth century. The great mine (Stora Kopparberg) at Falun became the single most important source of European copper. Falun had been worked since at least the eleventh century but before the 1600s the standard of mining and smelting fell significantly below German best practice. It was to remedy this historic backwardness that the Swedish state offered generous inducements to foreign entrepreneurs who would bring capital and expertise to Sweden. The policy had startling results between the 1620s and 1640s as industrialists from Germany, Wallonia, and the Dutch Republic relocated to the northern kingdom. They brought about an organizational transformation of Swedish production. Hitherto, Falun's copper had been exported in an unrefined state. The new arrangements were intended to keep the high value-added parts of the commodity chain (that is, the refining of copper and the manufacture of brass) within Sweden. Thus the Mommas, originally from Aachen, became prime movers in the new Swedish brass industry. Louis De Geer, a native of Liège, who is now better remembered for revolutionizing the Swedish iron industry, became proprietor of the important brass works at Norrköpping, while the Kocks, a Liègeois family like the De Geers, oversaw the refinery and mint at Avesta that processed Falun copper. Swedish hegemony was relatively short-lived, however. As the excavations at Stora Kopparberg were driven deeper the quality of the ore deteriorated. The workings also became more unstable. A sequence of underground cave-ins culminated in 1687 in a collapse so total that it halted production at Falun for several years thereafter. Sweden had bucked the trend in the seventeenth century, expanding while the German mining districts languished, but with the crisis of 1687 Sweden too regressed.

The eighteenth century saw a dramatic reordering of European copper making. For the preceding two and a half centuries the bulk of European copper had been smelted in the heart of the continent, within a band that stretched south from Scandinavia to the Alpine/Carpathian mountain ranges. In the eighteenth century everything changed. Most European copper came to be smelted on the continent's extremities. In the west, the Swansea District was ascendant; in the east, copper boomed along the Ural mountain range. Each knew stupendous growth in the eighteenth century. The Welsh copper sector, which scarcely existed at the start of the period, expanded at a phenomenal rate. Quantifying this growth is not easy, but we have fairly reliable output figures for British copper mines in 1771, with a suggested metallic content of 3,514 tons, most of which was smelted in the Swansea District.[7] Growth in Russia was no less impressive. A series of huge smelting works were developed beginning in the 1720s, first in the Urals and then the Altai Mountains; by the 1760s they were capable of producing 3,500 tons of smelted copper annually.[8] Both Russia and Britain represented a revolutionary departure, but their trajectories took radically different forms.

The expansion of Russian copper production was driven by a desire to break free of Swedish supplies. It was a matter of state, part of Peter the Great's determination to end Sweden's political and commercial hegemony in the Baltic. Russia would henceforth rely on its own resources and not add to the revenues collected by the Swedish treasury. As a form of import substitution, the policy was strikingly successful. Russian copper, however, afflicted by high production costs, was rarely competitive on wider European markets. The metal had to be absorbed by domestic users. Not the least of these was the state itself, which minted much of the output of the Ural works. Russian copper did not, therefore, greatly impinge on the global market. But British copper did. At the start of the eighteenth century, Britain's domestic market was still limited in extent, and the British state, unlike its Russian counterpart, was not willing to stockpile copper in the form of coinage. In consequence, British producers became heavily dependent on external markets in the Caribbean, West Africa, and India.

None of that could have been guessed at in the mid-seventeenth century, when Britain had no copper industry of its own to speak of. Most of the copper consumed in the British Isles originated in Sweden. That is not to say that the mining and smelting of copper had not been tried. On the contrary, state-sponsored attempts to promote an English copper industry had begun in the 1560s, just as the central European mining districts began to decline. The key figures were Germans already associated with copper mining in the Tyrol and Upper Hungary. The Germans were as keen to secure fresh sources of ore as the Crown was to end England's chronic dependence on foreign copper. As a result, a consortium of Augsburg merchants headed by Daniel Höchstetter secured monopoly rights over the mining of copper and precious metals in a selection of northern and western English counties and the whole of Wales. Together with some distinguished English investors, the Germans were incorporated as the Society of the Mines Royal in 1568. The Lake District had already been identified as a place of mineralogical promise, and the neighborhood of Keswick was to be the chief center of activity for the Society of the Mines Royal and its German workforce in the decades that followed.[9] A similar arrangement, combining Englishmen in high places (the shareholders included Elizabethan luminaries such as Sir William Cecil and the Earl of Leicester) with central European expertise, was put in place to stimulate brass production. The Society of the Mineral and Battery Works was granted the exclusive right to produce brass and wire in England. It established works at Tintern in the Wye Valley to do so.[10]

In the 1580s the Society of Mines Royal began to prospect for ores in the English southwest. The initial signs were so hopeful that the society set up smelting works at Aberdulais in the Neath Valley to process Cornish ores. This premonition of the later Swansea District was overseen by a veteran of the Keswick operation, Ulric Frosse, whose name betrays his German origins. It did not pay. Whether through a lack of capital or an inability to drain the Cornish mines effectively, the works at Aberdulais was allowed to lapse before the end of the sixteenth century. The parent works at Keswick was also struggling. It was maintained in operation until the outbreak of the civil wars of the 1640s but only as a losing concern. The brass-making efforts of the Society of the Mineral and Battery Works were equally unavailing. Because English copper smelting was so unreliable, brass manufacturers in England were forced to use Swedish imports as feedstock. Such dependency did not bode well; it left brass makers vulnerable to fluctuations in supply. Even well-connected foreigners, like Jacob Momma, a member of the famous Aachen dynasty, who introduced advanced brass-making techniques at Esher in Surrey in the 1640s, fell victim to shortages.

The privileges granted to the Society of Mines Royal did not lead to sustained copper production in the British Isles; nor did the Society of the Mineral and Battery Works yield much in the way of brass. Their deficiencies left England and Wales almost as dependent on continental imports in the mid-seventeenth century as they had been in the mid-sixteenth.[11] Indeed, it may be that the two Elizabethan monopolies actively stymied the emergence of an effective copper-smelting / manufacturing sector. That, at least, was the argument of those who pioneered the serious study of the copper trade in Britain. According to this historiographical tradition, the monopoly companies, unable to satisfy domestic demand themselves, hounded those who might, prosecuting private entrepreneurs who trespassed on their privileges. The restrictive grip of the Society of Mines Royal was identified by Henry Hamilton in his classic history of the industry as "a stumbling block to private enterprise" in the 1600s and "the main cause of the decay of the brass and copper industries in the second half of the century."[12] The monopoly rights granted to the society were maintained by successive monarchs until the Revolution of 1688 toppled the Stuart dynasty. Only then did the deregulation of the industry become politically viable. The Mines Royal Act of 1689 annulled the Society's legal stranglehold over copper mining and opened the way for the free play of private investment.

Yet there is reason to doubt this account. J. R. Harris, in his introduction to the 1967 reissue of Hamilton's work, suggested that the "abolition of monopoly was probably the formal recognition of a changed economic and technical situation . . . rather than an important liberating force" in its own right.[13] George Hammersley, writing in the early 1970s, was equally skeptical about the ability of the Elizabethan monopolies to enforce their privileges. They were, he thought, ineffective parasites. The Society of Mines Royal, Hammersley noted, was usually content to license independent mineral prospectors for a nominal fee. "There is no sign," he concluded, "that, at any time, an enterprising man was deflected from the search for or the exploitation of copper by the company's resistance."[14] Hammersley attributed the enfeebled state of English copper smelting in the seventeenth century to a lack of demand. The market for copper and brass wares in England was simply too limited to support a domestic mining/smelting sector, especially if low-cost Swedish imports were available.

Yet if demand was so weak, for so long, how can the sudden resurgence in British copper smelting at the end of the seventeenth century be explained? The abruptness with which new capacity appeared in the decade and a half after 1690, when five new copper works and four brass works made their debuts, is striking. Was the emergence of a British copper industry a response to a failure of overseas supply? It is true that the Swedish copper sector was in difficulties. Output at Falun had been slipping even before the cave-in of 1687. The mine was eventually reopened, but mining operations were permanently impaired; Falun never recaptured its seventeenth-century heights. The crisis of Swedish copper must have resulted in a certain amount of unsated demand in the British Isles. Even so, imported copper continued to be of critical importance to the British market in the first decades of the eighteenth century. Swedish imports may have dwindled, but Sweden was not the only source of copper. Copper and cupreous products continued to arrive from Germany and Holland, and imports of copper from North Africa boomed in the 1720s and 1730s.[15]

Alternatively, is the rise of copper smelting in Augustan England to be explained by sharply rising demand rather than problems with overseas supply? After drifting downward in the last decades of the seventeenth century, the population of England and Wales was on an upward course by the second quarter of the eighteenth, and with it the market for copper and brass. Quantifying domestic demand is by no means easy, but it seems unlikely that home consumption could have increased at such a rate as to justify the rash of new works in the Bristol region, then the Swansea District, in the early decades of the eighteenth century. For that to be explained, we must look to a wider Atlantic context.[16]

Copper in the Atlantic Economy

Domestic demand was buoyant, but the demand for copper in colonial markets was increasing at an exponential rate. The population of England's Caribbean colonies was rocketing, with enslaved Africans being landed in prodigious numbers.[17] The captive population was dedicated to a peculiarly concentrated form of agro-industrial production, the growing and processing of sugar cane—an industry that required large volumes of copper for its success. Sugar was made by boiling the sticky liquor obtained by crushing cane. This was done in receptacles known, for good and obvious reasons, as "coppers."[18] These vessels were encased in a boiling house, a substantial stone structure built on two levels. Below, teams of slaves loaded bagasse (dried-out trash from the cane fields) into the furnace that kept the superincumbent coppers bubbling. On the upper floor another set of slaves tended the copper pans, four or five in number, stirring the liquor and skimming off the dirt and residual mill debris that was brought to the surface.

As evaporation took hold the liquor thickened. When it reached the critical consistency it was ready to be ladled into an adjacent, somewhat smaller copper. Another boil and further concentration of the liquor ensued; then another after that in a still smaller copper vessel. Finally, the sticky syrup was decanted into the "tache," or "teach," the smallest copper, in which the sugar was granulated. The coolers in which the warm sugar was allowed to settle were also made of copper, as were the skimmers and ladles used by boiling-house slaves. How much metallic copper then did the boiling house embody? Working with a sample of leases and probate inventories from Barbados and Jamaica in the years 1670–1700, Nuala Zahedieh has estimated that a boiler was required for every nine slaves engaged in sugar production and that an average copper would have weighed 2.2 hundredweight. This ratio allows a rough calculation of the copper used in sugar manufacture to be made. The 115,000 captive laborers on the English islands in 1700 "would have needed around 8,517 copper boilers and other equipment weighing altogether an estimated 1,123 tons, almost twice the weight of England's copper coinage at the time."[19] And as sugar production in the islands grew, so did the volume of copper consumed. By 1770 there were 434,000 enslaved workers in the British Caribbean, implying a stock of 32,148 boiling coppers. The total weight of embodied copper—including boilers, coolers, and boiling-house implements—would have come to 4,243 tons.[20]

This is likely to be an underestimate. Zahedieh takes the largest copper in a boiling house suite, the clarifier, to have been of 180-gallon capacity and the average tache to have held 30 gallons. A suite of three boilers and a tache amounted to nearly half a ton of copper. This may well have been standard for the late seventeenth century, but on the eve of the American Revolution clarifiers of 300- or even 400-gallon capacity were in use. The Jamaican planter Florentius Vassall ordered one of 400 gallons in 1774. The pan was to be seven feet in diameter at its brim, narrowing to five feet at its base; the copper was to be as "thin as possible & well & smoothly polished."[21] And as clarifiers went, so did taches. The London manufacturing coppersmiths George & William Forbes, who furnished Vassall with his 400-gallon clarifier in 1774, received orders for taches of 70 or 80 gallons in the same year.[22] The weight of such vessels was inevitably much greater than their equivalents of a hundred years earlier. A consignment of coppers (two clarifiers and four taches) shipped to Jamaica by John Freeman & Co. of Swansea in 1790 weighed 1.9 tons, suggesting that an "average" copper would have been 7.9 hundredweight—a massive advance on the 2.2 hundredweight estimated by Zahedieh for the late seventeenth century.[23] The 32,148 coppers thought to have been present in the British islands in 1770 would therefore have weighed in at 12,845 tons. Adding an additional 20 percent to allow for the weight of coolers and other cupreous equipment gives a total in excess of 15,400 tons, a figure that is more than three and a half times that suggested by Zahedieh.[24]

Caribbean demand was not a constant, admittedly. Sugar was an industry that boomed and slumped, with the slumps usually brought on by wartime privateering that choked off the supply of enslaved labor and other necessities to the islands. The impact of war is clearly revealed in the archive of the coppersmithing business of George & William Forbes. Orders from the West Indies dried up in 1776 as the American privateers began to swarm into Atlantic sea lanes. The Forbes brothers were saved from ruin by switching to military production, furnishing the Royal Navy with very large volumes of copper sheathing. In more favorable conditions, however, Caribbean demand would reassert itself. It did so massively at the turn of the nineteenth century, when the collapse of France's premier sugar colony, Saint-Domingue, allowed British planters to make huge profits and when British naval supremacy allowed for the seizure of territories like Trinidad or Demerara that were well suited to sugar production. The effects are visible in the history of another firm of London coppersmiths, John Grenfell & Co., which had one riverside manufactory at Battersea and another on Ratcliff Highway to the east of the City. A list of debts owed to John Grenfell & Co. when it became insolvent in 1803 indicates that the firm's most important clients were merchants who shipped equipment and provisions to planters.[25] Indeed, the most indebted of John Grenfell & Co.'s customers were London's wealthiest West India houses: Boddington & Sharp, Wedderburn & Co., and the Hibberts.[26] They were responsible for conveying large volumes of copper to boiling houses in the New World.

The boiling house was not the end of it, though, for the boiling process had two outcomes. One was muscovado, as the raw brown sugar was known; the other was molasses, the syrup that was drained from the muscovado. Molasses was a by-product rather than waste, for there was good money to be realized on it as the base material for rum. Indeed, there was an Atlantic-wide boom in distilling in the late seventeenth and eighteenth centuries and a corresponding boom in demand for copper distilling equipment.[27] Part of this demand came from the West Indies, from planters who wanted to distill their own molasses. An even more substantial demand came from British North

America. The continental colonies supplied the islands with lumber and provisions, of which the islands, cleared of their timber and crowded with slaves, were in sore need. Mainland merchants took molasses in payment and built rum distilleries to make profitable use of it. As a result there were 140 distilleries in North America by 1770, most of them in New England, the source of many of the barrel staves and most of the animal protein upon which the Caribbean sugar sector and its enslaved workforce depended.[28] There were also distilleries in Britain. London had 63 of them on the eve of the American Revolution.[29] All of these distilleries, whether in the Caribbean, New England, or London, were supplied with stills by British coppersmiths. They came in all sizes. Stills made for plantation use, a Jamaican planter reckoned, varied "greatly in point of size and expence, according to the fancy of the proprietor, or the magnitude of the property."[30] The largest could hold from one thousand to three thousand gallons of liquor. Those used in London were larger still, and grew rapidly over the course of the eighteenth century as the industry became more heavily capitalized. Cucurbits of five thousand gallons were in use in the capital by the 1750s, and by the end of the century titans of ten thousand gallons were at work.[31]

The refining of sugar—the conversion of soft brown muscovado into hard-baked white loaves—used copper too. Sugar pans, shallow and flat-bottomed, were usually about four feet across. Four were required at each boiling house: "two clarifying pans, one skimming vessel and one boiling pan."[32] In the sixteenth and seventeenth centuries the leading centers of European refining were to be found in the Low Countries and Germany, but by the 1750s Britain had come to the fore: "There are about eighty refining Houses in and around *London*, and twenty at least at *Bristol*; and it is very well known that there are likewise refining Houses at *Chester*, *Liverpool*, *Lancaster* and *Whitehaven*, at *Newcastle*, *Hull* and *Southampton*, and some in *Scotland*; I think there can hardly be fewer than 120 in all."[33]

The Atlantic sugar complex constituted a major new source of demand for British copper. Every facet of the business—sugar boiling, sugar refining, and the distilling of rum—called for copper. Copper featured in one other part of the transatlantic sugar complex too, for copper and brass articles played a key role in the acquisition of the plantation world's enslaved workforce. Copper had been an important part of material culture in West Africa for many centuries. Although the archaeological evidence is scant, the signs are that sub-Saharan smelting can be dated to the first millennium BCE. Copper-bearing minerals are rare in West Africa, however, so the region drew on copper-producing areas to the north of the Sahara.[34] Yet there was a limit to what could be brought in by trans-Saharan caravan, so when Portuguese ships first appeared on the Guinea coast in the fifteenth century with cuprous trade goods on board they found a ready market awaiting them.[35]

The Portuguese sourced copper and brass in northern Europe. Battery wares from Stolberg and Aachen were available at Dutch ports, while copper from central Europe could be had at Hamburg or Antwerp.[36] When the Dutch began to make inroads on Atlantic slave trading in the seventeenth century they adapted this supply chain to their own purposes. Brass wares continued to flow in from northern Germany, but copper was now shipped in from Stockholm during this, the heyday of Stora Kopparberg, as well as from Hamburg. When the English took to slaving in earnest in the closing decades of the seventeenth century, they followed suit. Brass was purchased in Holland; copper came from Sweden.[37] The Royal African Company, which exercised a legal monopoly over English trade with Africa until 1698, was always likely to remain loyal to Baltic suppliers, largely because the merchants most closely involved in furnishing Swedish copper were themselves shareholders in the company and, indeed, very often among its leading officers. When the Royal African Company's monopoly rights were rescinded, however, the field was opened up for traders who felt no loyalty to merchant houses in Stockholm or Hamburg.

The independent traders who flooded into English slaving after 1698 constituted another important pulse of new demand for copper and brass. This demand was held in check by wartime conditions in the first decade and a half of the eighteenth century but in the postwar years its force was felt for the first time. Most of this growth was accounted for by sailings from Bristol rather than London, where the Royal African Company was headquartered. The new copper and brass works that sprouted around Bristol in the early eighteenth century were perfectly placed to serve the city's slaving fleet. Indeed, there was an easy symbiosis between the emergent copper industry and the trafficking of slaves. Many Bristol merchants, like Abraham Elton (1654–1728), invested in both because copper was such a staple element of the Guinea trade. Lengths of copper ("Guinea rods") acted as a currency in many slave marts, as did manillas (bracelets formed from brass or a copper-lead alloy). These were churned out in large quantities, primarily from the brass and copper works that encircled Bristol, but the making of trade goods for slave merchants was also a feature of pioneering works in the Swansea District. The proprietors of Llangyfelach, the first copper smelter in the Tawe

Valley, built a water-powered battery mill at Forest at the end of the 1720s, which could be used to manufacture Guinea rods, while the works at White Rock, erected in the 1730s, featured a "manilla house."[38] It was a lucrative trade, all the way to British abolition in 1807. The invoice of the *Africa*, which cleared Bristol for the Bight of Biafra in 1774, is indicative. It included four thousand copper rods, two hundred "Neptunes" (shallow brass bowls over a meter across), and sixteen casks of manillas. There was well over four tons of copper here, in pure or alloyed form.[39]

All in all, the Atlantic slave-sugar complex constituted a potent new source of demand, one that was of decisive importance in the first half of the eighteenth century. It was followed, as we shall see, by other landscape-changing pulses of external demand: the purchasing of copper for the South Asian market by the East India Company from the 1730s; then, in the 1770s and 1780s, naval demand for copper sheathing. Yet that demand could never have been satisfied without a technological transformation of copper smelting. Producing copper in the British Isles by means of the charcoal-fueled methods that prevailed in central and northern Europe would have made unsustainable demands on local energy reserves. Copper could be smelted in the British Isles using vegetable fuel but not in large volumes, certainly not in the volumes being produced at Swansea by the second half of the eighteenth century. The growth rates registered by the copper industry in England (and increasingly in Wales) in the eighteenth century required a different energy source: coal.

Coal was to be fundamental to Swansea Copper, as it was to be to British society as a whole. Careful coppicing and close management allowed early modern English landowners to maximize the yield of firewood and charcoal from their estates. Even so, the energy at their disposal was finite. It could not be otherwise on an archipelago that was by wider European standards quite lightly wooded. Moreover, population growth and the prosperity of England's staple woolens industry demanded that land was put to arable and pastoral uses. Grain and fleeces took priority over stands of timber. Energy shortages were avoided, however, by tapping sources that were subterranean and seemingly limitless. Being blessed with an abundance of coal, the British were able to burn fossilized biomass rather than harvest living matter. Not only were the British Isles generously endowed with coal deposits; those deposits were often accessible to river traffic and seagoing craft. The Great Northern Coalfield of Durham and Northumberland pioneered the British model of seaborne energy provision. From the middle of the sixteenth century, collieries along the Tyne and Wear began to ship coal to London in large volumes. The coal trade continued to boom in the seventeenth century. Indeed, the heaving growth of London, from 200,000 inhabitants in 1600 to 575,000 a hundred years later, would have been unthinkable without a corresponding expansion of mining in the Northeast: "Whereas when we are in London and see the prodigious fleets of ships which come constantly in with coals for this increasing city, we are apt to wonder whence they come and that they do not bring the entire country away; so on the contrary, when in this country we see the prodigious heaps, I may say mountains of coals which are dug up at every pitt, and how many of these pitts there are, we are filled with equal wonder to consider where the people live that can consume them."[40]

The coal trade along England's east coast set a pattern. Coalfields that were coastal or served by navigable rivers were well placed to supply similarly situated urban markets. Thus, collieries multiplied in seventeenth-century Cumberland to feed the lucrative Dublin market, and landowners on the Shropshire coalfield, which was bisected by the Severn, were able to sell coal upriver and down.[41] The ability to serve markets around the Bristol Channel was a critical stimulus to the development of a coal industry in southwest Wales in the seventeenth century. Coal measures cut across the Neath and Tawe Valleys just a mile or two from the coast. Seagoing craft could float upstream on the tide, tie up at a riverside staithe, load with coal, and then drift back down to the river's mouth as the tide ebbed.[42] This was a substantial trade, one that was well established before the coming of the copper industry. Neath shipped coal primarily to Somerset. Swansea's shipments were focused on Devon and Cornwall, the emergent naval base at Plymouth, and ports along Ireland's southern coast.[43] (By way of contrast, the coal reserves of the landlocked Rhondda Valleys, less than ten miles as the crow flies from Neath, remained undisturbed until the railroad made them accessible in the 1840s.)

New Coal Technologies

This was an environment in which the coal-fired reverberatory furnace (figure 1.1), the basis for the Welsh Process of copper smelting, could flourish. The Welsh Process did not originate in Wales, however; nor was it initially deployed to smelt copper. Indeed, there was nothing about the reverberatory furnace that mandated the use of coal. The defining principle of the reverberatory furnace was that the fuel being burned and the materials upon which the heat and combustion gases were to act were kept separate. Furnaces of

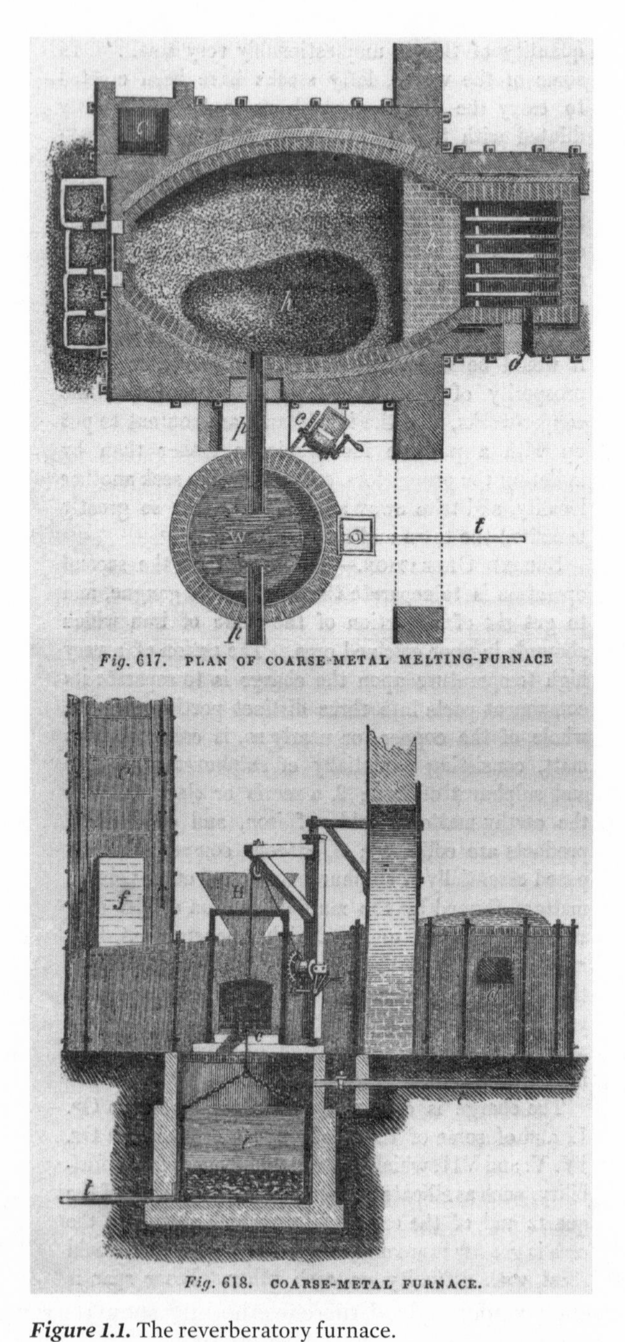

Figure 1.1. The reverberatory furnace.
The furnace had three essential features, shown here from top to bottom in plan view (Fig. 617) and side view (Fig. 618): (i) the fire-grate in which coal was burned, (ii) the hearth in

that sort, known at first as cupolas, were used by medieval bell founders to melt their metal. By the start of the sixteenth century they were employed by gun founders in both Italy and Germany to melt bronze, consuming wood or charcoal as fuel. The use of mineral coal cannot definitely be attested to before the seventeenth century, when the English Crown granted a patent to one John Rovenson for a new method of smelting. Rovenson's description of his apparatus in *The Treatise of Metallica* (1613) is clearly that of a reverberatory furnace, "wherein the metall or material to be melted or wrought is kept divided from the fewell."[44] Rovenson added that such furnaces could be constructed "without bellowes," another of the hallmarks of the reverberatory method. Combustion was intensified by a very strong chimney draft that drew flames from the fireplace over the hearth in which the "material to be melted or wrought" was seated. The ceiling of the hearth was inclined downward so as to deflect (or "reverberate") heat directly onto the furnace charge.

There is no evidence that Rovenson actually put such a furnace into production, but the principle of coal-fueled reverberatory furnaces is found in a

---

which ore was smelted, and (iii) a flue ("*f*" in the side view) leading to the stack. The height and narrowness of the chimney stack created a powerful draft that swept flame and combustion gases from the fire grate across the hearth. The sloping roof of the hearth served to deflect (or "reverberate") radiant heat down onto the mineral charge.

The furnace shown here was used for the second operation in what had by the nineteenth century evolved into a ten-part process. Calcined ore was deposited in the furnace via the hopper marked "*H*" in the side view. The smelting process yielded a liquid copper matte that gathered in a depression in the hearth floor ("*h*" in the plan). This was tapped into a tank of water ("*W*" in the plan and shown in section in the side view). The matte granulated in contact with the water. Winched clear of the tank in baskets, it was dried and carted off to a different furnace for the third operation.

Cyclopaedia of Useful Arts, Mechanical and Chemical, Manufactures, Mining and Engineering, vol. 1, *Abattoir to Hair-Pencils*, ed. Charles Tomlinson (London: George Virtue, 1852), 432 number of patents and prospectuses of the mid-seventeenth century. They were first put to serious use in the lead industry, which was a much more significant industrial sector than copper in the seventeenth century.[45] Indeed, the first reverberatory furnace to be built in the Swansea District was intended to smelt lead quite as much as copper.[46] Smelting lead was certainly the aim of the earliest pioneers. It was Sir Clement Clerke (d. 1693), a Leicestershire landowner, who appears to have done most of the development work.[47] "Cupoloes for reduceing Lead-Ore with Sea-Cole," the natural historian Dr. John Woodward claimed, "are an English Invention . . . contrived by that most excellent Mineralist Sr Clem^t^ Clark."[48] However, when a patent "to melt and refine lead ore in close or reverberated furnaces with pit-coal" was issued in 1678 it went not to Clerke but to George Villiers, fourth Viscount Grandison, who had been sponsoring experimental smelts of his own. In all likelihood neither Grandison nor the Clerkes had yet settled on a reliable method of smelting lead ore; a pooling of expertise seemed the best way forward. Collaboration between Grandison and the Clerke family led to the establishment of the Bristol Lead Company and the building of the Leigh Woods smelting works on the left bank of the River Avon, just below Bristol.[49] By the early 1680s, however, Sir Clement and his son Talbot Clerke, who had active management of the works, had fallen out with Grandison and quit the Bristol Lead Company's works. Proceedings in the Court of Chancery ensued, whereby Grandison succeeded in having the Clerkes debarred from smelting lead independently. Confident in their mastery of the reverberatory furnace but excluded from the lead business, the Clerkes looked for another outlet for their expertise. They turned to copper, obtaining a patent in 1687 that granted them the same sort of privileges in copper smelting that Grandison enjoyed over lead.

The Clerkes' early efforts with copper were conducted at Putney, to the west of London, where they erected a copper-smelting furnace "which was the Pattern & Original of all the rest."[50] A full-scale smelting operation followed at Rownham Mead on the edge of Bristol, provocatively close to the Bristol Lead Company's works. It was undercapitalized and folded in the early 1690s. Nevertheless, the viability of the reverberatory furnace as a means of smelting copper had been clearly demonstrated. Others took notice. Imitations of the Rownham Mead operation sprang up in a variety of settings. Many were ephemeral, without the vigor to survive into the eighteenth century. John Woodward itemized short-lived reverberatory works at "White-Haven in Cumberland: near Ashburton, in the Peak: near Auderly

Edge, Cheshire: near Dizart in Flintshire: at Neath in Glamorganshire: near St Austils, in Cornwall: near the Saw-Mill, in Southwark: [and] at Fox-Hall [Vauxhall]."

Others were more enduring. Writing in the early eighteenth century, Woodward mentioned two on the River Wye, a tributary of the Severn that marked the border between Wales and England: Upper Redbrook (est. 1690) and Lower Redbrook (est. 1691). Two more were to be found along the River Avon between Bristol and Bath: Conham (est. 1695) and Crews Hole (est. 1706). Another was at Neath (est. 1694), which offered a foretaste of where locational logic was soon to concentrate copper production in Britain. Finally, there was "Mr Morgans [works] at St Ives in Cornwall."[51] These were works that had moved beyond the experimental. Each featured several reverberatories because "Furnaces of different Structure" had to be used sequentially in order to reduce chemically complex ores. "They have commonly," Woodward wrote, "Melting-Furnaces, Roasting or Calcineing-Furnaces, & Refineing-Furnaces, in the same work." There was another common feature. They were all within striking distance of the Bristol Channel. They were all therefore able to receive shipments of Cornish ore. "They fetch all their materials from Cornwal and Devonshire," it was reported from Conham in 1697; "they find it cheaper to bring the Mine to the Cole, than to carry this to ye mine."[52]

Equally, all of them were able to access coal easily, either because they were immediately adjacent to coal measures or, in the case of the furnace in Cornwall, because seaborne coal from South Wales was readily available. The region defined by the Bristol Channel had clearly been identified as a very good place to smelt copper. The question to be resolved was where, within that broad region, was optimal?

Swansea's Apprenticeship, c. 1690–1750

If the development of coal-fired smelting was the technological breakthrough that sparked the revival of copper smelting and refining in Britain, the question of where this new method could be worked most profitably took some three decades to answer. Place and process were inextricably linked. The new generation of smelters knew this, but it was by no means clear in the period from the 1690s to the end of the 1720s where the reverberatory process could best prosper. Given the volumes of coal needed to fuel the new process, the availability of cheap coal has been identified by historians as the factor that gave South Wales the competitive edge over rival smelting locations.[1] Coal mining around Swansea was already well developed by the beginning of the eighteenth century, as we have seen. Moreover, the type of coal being mined in the region was particularly effective as a furnace fuel. The ability of the early copper smelters in the region to exploit this resource was highly significant in securing the success of copper smelting in the district, but it was not the only factor. A web of local, regional, and international conditions combined to determine the cost effectiveness of copper smelting in Swansea. The securing of adequate supplies of copper ore in an era when output from Britain's major ore field, in Cornwall, was rather patchy, was one major challenge. The development of robust financial arrangements, linking the smelters with merchant capitalists who could access new overseas markets was another. The Swansea District's emergence as a copper-smelting location in the early decades of the eighteenth century depended on its ability to overcome both of these hurdles. By the 1750s it had demonstrably done so: the apprentice had become the master.

In the opening years of the eighteenth century, however, visitors to Swansea or Neath would have found little to presage the future development of the district as Britain's copper-smelting location par excellence. The Bristol region was a more plausible location, one to which the early copper masters gravitated. It had local coalfields at Kingswood, just outside the city, and in the Forest of Dean, immediately to the north of the Severn. Hence the foundation of copper works at Upper Redbrook and Lower Redbrook in 1691 and 1692 respectively; they could draw upon the Dean coal measures and receive Cornish ores via the Wye. Hence too the building of smelting works at Conham (1695) and Crews Hole (c. 1706) on the Avon, sites that were sandwiched between the river and the Kingswood outcrop. Positioning works around Bristol was an attractive option because the city itself was a major market. At the start of the eighteenth century, Bristol vied with Norwich to be England's second city. It was certainly the second-ranked English port, one with a leading role in England's Atlantic commerce. Importantly, it was also ideally placed to be a center of brass manufacture. Calamine (zinc carbonate) was mined in the Mendip Hills, a limestone ridge to the south of the city. Battery mills to work up brass had traditionally clustered in the southeast of England, drawing on tributaries of the Thames for water power and using imported brass ingots as feedstock. Now, brass began to be manufactured in the Bristol region, using locally smelted copper and Mendip calamine, and exploiting the power of the Avon and its tributaries. Brass production began at Baptist Mills, just east of Bristol, in 1702 and at Keynsham on the Avon shortly afterward. A penumbra of rolling and battery mills followed, beginning with Swineford, upstream from Keynsham, in 1708. Over the next few years processing mills were built at Saltford, also in the Avon Valley, and at sites along the River Chew, which tumbled down from the Mendip Hills.[2]

The copper/brass enterprises that sprouted around Bristol in the first years of the eighteenth century were closely integrated. Indeed, they were often in shared ownership. The partnership that smelted copper at Upper Redbrook, for example, also controlled the mills at Swineford and at Pensford in the Chew Valley. The Bristol Brass Company lay at the heart of a still more extensive network. It combined the copper works at Conham and Crews Hole with the brass works at Baptist Mills and three battery works in Bristol's hinterland. The early Bristol copper companies were united in one other thing: their reliance on a handful of technical adjutants who played a key role in stabilizing the Welsh Process. Gabriel Wayne was one of these. He was "principal manager" at the Clerke family's Rownham Mead works in the 1680s. Wayne then, at the start of the 1690s, set up the English Copper Company's smelting works at Lower Redbrook. After that, in the mid-1690s, he established the works at Conham in partnership with the Bristol merchant Abraham Elton. Members of the Coster family were another ubiquitous presence in the nascent Bristol copper trade. The Costers originated in the Forest of Dean, where they had been active in the iron industry, but they were drawn into the experimental smelting of lead and copper in reverberatories that blossomed around Bristol from the 1670s. John Coster (1647–1718) came to prominence at the Clerkes' copper-smelting operation at Rownham Mead alongside Gabriel Wayne. He then superintended the works at Upper Redbrook on behalf of a consortium of London merchants marshaled by William Dockwra. John Coster played a critical role, contemporaries agreed, in adapting the reverberatory furnace to the smelting of copper ores. His significance extended beyond that, however. As will be detailed below, John Coster became a major force in the marketing of Cornish ores.

Nonferrous Smelting Enterprises at Neath

Innovation in the smelting of copper centered on Bristol; by comparison, the Swansea District (figure 2.1) was a slow starter. Early developments revolved around the town of Neath, which had a long tradition of metal smelting dating back to the Elizabethan period. In the mid-1690s two new smelting ventures emerged there which, despite their close geographical proximity and contemporaneity, used different approaches to the task of copper smelting. The first, in 1693, was located on an old iron smelting site at Neath Abbey, which was leased and adapted for copper smelting and refining by Benjamin Gyles, Elizabeth Hoby, and John Champion.[3] There was an associated mill upriver at Cwmfelin, where an array of water-powered battery machinery was in place.[4] At this site, coal was available locally, but the smelting operation, like those in the Bristol region, was dependent on Cornish ores brought in by sea. Shipments arrived in 1696 and 1697, and the copper produced at the works was sent to buyers in Bristol.[5] This Bristol connection may have been the key to the takeover of the site in 1706 by a new partnership comprising a Bristol surgeon, Dr. John Lane; John Pollard, of Redruth, Cornwall, who was a relative of Lane's and a Cornish landowner with mining interests; and a third partner, Thomas Collins.[6] This proved a relatively short-lived grouping that came to an end in 1716.[7] It is likely that the Neath partners found it difficult to compete with their counterparts located closer to the main brass markets. Copper smelting at Neath Abbey continued in the short term with the Welsh Copper Company at the helm. It sounded promising, but it did not prosper. Indifferent business management by the Welsh Copper Company,

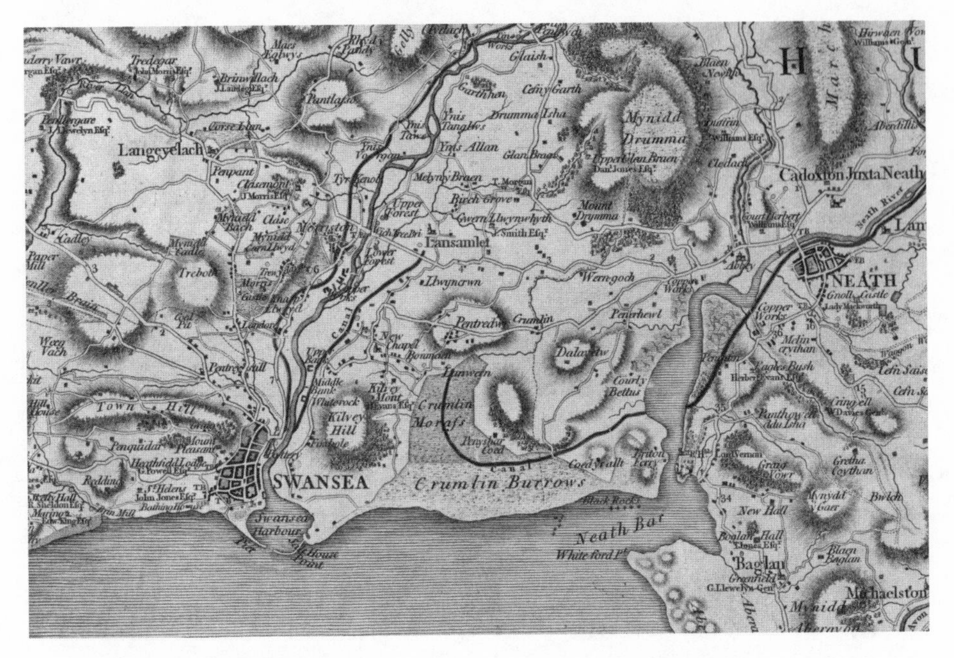

Figure 2.1. Glamorgan 1799.
Extract from the survey of Glamorgan undertaken by George Yates and published by John Carey of London. It shows the location of the early copper works on the Neath and Swansea Rivers.
West Glamorgan Archive Service, D/D Z 134

whose role in the copper industry was described by one authority as "relatively undistinguished," combined with the volatile financial climate of the 1720s, was probably to blame for the demise of their Neath Abbey venture.[8]

The second Neath smelting operation was at a site named Melincryddan, on the River Neath, where Humphry Mackworth built smelting furnaces in 1695. Mackworth was from Shropshire, on the English/Welsh border, the second son of a family of landowners, whose industrial interests in South Wales were enhanced following his marriage in 1686 to Mary Evans, daughter and heir of Sir Herbert Evans of the Gnoll Estate in Neath.[9] His new smelting site at Melincryddan had much in common with the Neath Abbey works in terms of location and convenient access to coal, but a key difference was the fact that Mackworth owned his own ore mines in Cardiganshire. He attracted investment through the Company of Mine Adventurers and sought to establish an integrated Welsh-based network linking ore mines, collieries, and furnaces. Such an ambitious and self-sustaining business plan was conceived, according to one historian, not just for personal profit but also as a way of alleviating poverty in the region and delivering on a religiously motivated duty to succeed in business.[10]

Humphry Mackworth has been described as "the outstanding figure" in the industrial rejuvenation of late seventeenth-century South Wales, but his role in establishing the Swansea District as a copper-smelting center was somewhat indirect.[11] Initially, his Melincryddan operation seemed to pose little threat to the dominance of the Bristol and Wye Valley firms because his efforts there were focused on lead smelting. Lead was being used for a variety of purposes in this period, as a roofing material and in the production of pipework and other items for the construction trade.[12] Moreover there were a number of lucrative spin-offs from the lead-smelting process. One was the production of litharge, or lead oxide, which was sought after by druggists, surgeons, and glass manufacturers, among others, in the early eighteenth century.[13] Another was the extraction of silver from lead ore. This was shipped to the London mint for coinage. Much of the produce of Mackworth's mid-Wales mines consisted of lead ore, which he shipped direct to the furnaces at Melincryddan. With such a heavy cargo, the cost of carriage loomed large and the importance of being able to unload the ore as close to the smelting site as possible was paramount. Mackworth's smelting works was less than half a mile from the River Neath, but even this relatively short distance did not eliminate the need for investment in transport infrastructure. In 1698 he built a 300-yard waterway from the River Neath toward Melincryddan so that "small vessels with thirty ton might come up within four hundred yards to the smelting and refining houses."[14] In 1703, he was empowered by his Company of Mine Adventurers to purchase "troughs," or barge-like vessels, "fit for loading and unloading vessels that can't come up the river."[15] Although these initiatives, designed to improve the cost effectiveness of lead smelting at Melincryddan, did not directly impact the copper trade, they demonstrated how carriage costs could be reduced and they magnified the significance of the navigability of waterways as well as the waterside location of smelting works.

Lead-smelting developments helped prepare the way for the copper smelters in other respects too. It was through the process of building up their lead-smelting operations at Neath that Humphry Mackworth and his partners began to amass a labor force with up-to-date knowledge of smelting processes. Knowledge transfer between the two industries had already shaped the smelting quarters on the Rivers Avon and Wye, where Clerke, Coster, and others had honed their metal-refining skills at the lead furnaces. In the Swansea District, similar patterns applied. Key lead-smelting expertise was brought into the district from elsewhere. Given the newness of the coal-fired reverberatory furnace method of smelting copper, knowledge of techniques was limited, and a background in lead smelting was as good a route as any into the new industry. Robert Lydell, Mackworth's chief refiner, recruited for the opening of Melincryddan works in 1698, had worked at Flint in North Wales, and at Newcastle, before his services were sought at Neath.[16] His metallurgical skills became known after he patented a process for separating silver from lead in the refining furnace.[17] Additional workers had to be recruited locally and trained in the appropriate skills. Around forty workers were needed to operate the works at Melincryddan. Records dating from around 1700 show how this workforce was distributed: twelve men to work six melting furnaces, a further sixteen men to work the calciners, two to operate the stampers, two to make bricks, and six to do general laboring work. Added to this were listed a mason, a smith, and a carpenter.[18] Labor shortages were resolved by recruiting young, local laborers, presumably with little or no prior knowledge of metal-smelting techniques, to bolster the ranks of the trained furnacemen and refiners. In February 1699, for example, Robert Lydell wrote to Mackworth warning that "we have not workmen enough to refine and smelt both at a time . . . if we had three or four more although but boys they would then keep two smelting furnaces and refining going. . . . If you think it proper to take any more I think these four following: Charles Evan, William Lewis, Edward Rogers and Evan Rees who seem to be active willing young fellows would be most proper. . . . These men have been all labourers in the building and have behaved themselves very well."[19] By the 1740s, the pool of smelting expertise available locally was recognized as one of the strengths of the district as a smelting location. A Mackworth agent, albeit in a sales pitch to a prospective lessee of Melincryddan, claimed that "as good common workmen as most in the kingdom for calcining and melting of copper" could be found there.[20]

The phase of lead smelting at Melincryddan not only helped in the process of building up a core of workers with knowledge of the smelting process; it also helped demonstrate the suitability of the coal, mined locally in the region, for use in the fuel-intensive metal-smelting process. Neath and neighboring Swansea were situated in the midst of a coalfield that was already being mined extensively by landowners and capitalist investors. Wales was not the most productive coalfield region in the British Isles in the late seventeenth and early eighteenth century. Estimated annual output from the mines in North and South Wales of two hundred thousand tons by the 1680s was dwarfed by the scale of production in the Durham and Northumberland coalfield, which was yielding over a million tons per year in the same decade, but the rate of expansion it had experienced since the mid-sixteenth century was equal to, or greater than, these larger mining regions.[21] Within Wales, however, it was the western end of the South Wales coalfield, around Swansea and Neath and further west into Carmarthenshire, where coal mining was most advanced in the sixteenth and seventeenth centuries. Good-quality coal could be reached relatively easily via drift mines dug into the hillside or vertical shafts of up to 25 fathoms (150 feet); in the northeast of England, by comparison, there were mines of 300-foot depth by 1700.[22] The Mansel family was one powerful landowning presence in the Neath coal trade. Thomas Mansel had mines at Llansamlet and Briton Ferry that were yielding profits of £800–£900 per year in the period from 1706 to 1730.[23] The Evans family of Gnoll had also leased mines and extracted coal for export from Neath since the late sixteenth century, and their assets were acquired by Humphry Mackworth in 1686 through his marriage to the sole heiress of the Gnoll estate.[24] Mackworth produced more coal than he required for his own purposes at the smelting works. The correspondence of his chief smelter, Robert Lydell, indicated that he had excess available to sell to other customers: "I doe humbly conceive it would be of very great advantages to you to buy two or three vessels of about 80 tunns each yourself for then you might certainly vend your own coles in Cornwall, Ireland and other places and bring lead oare back . . . [when] other people and places came to know the goodness of them and what great quantities you could supply them with they would be certain to come themselves and trade with you at Neath."[25] The net result was that early copper-smelting ventures helped stimulate further growth in what was already a dynamic part of the British coalfield, with Nef estimating a quadrupling of output from the mining districts of Wales in the century after 1681.[26]

While the need for cheap and effective fuel led a number of the early Welsh smelters to enter into coal-mining ventures, they showed little inclination to invest in copper-mining ventures in Cornwall. There were considerable disincentives. Until the end of the seventeenth century, Cornwall's principal mineral product was tin, the mining of which was subject to Stannary regulations dating back several centuries. Under Stannary law, landowners had no automatic right to revenues from tin mined on unenclosed land, so there was no incentive for them to invest in the industry. Tin occurred at relatively shallow depths, but by the late seventeenth century tin miners were beginning to encounter copper lodes as they drove their workings deeper. The question arose as to whether copper was governed by the same Stannary regulations as tin. A 1683 lawsuit brought by Hugh Boscawen, a landowner in the copper-rich Chacewater district of Cornwall, resolved the issue: copper was *not* subject to Stannary law.[27] This ruling had far-reaching implications for the development of Cornish copper mining. The way was clear for landowners to become major players in the development of the industry, either by undertaking to mine on their own account, or by selling shares in their mines to interested investors.

This opportunity was timely, given the increasing demand for ore that accompanied the surge in investment in smelting works in the Bristol region in the 1690s. Indeed, the entrepreneurs most closely involved in the spread of reverberatory furnaces in the Wye and Avon Valleys were keenly interested in acquiring Cornish ores and in improving the performance of Cornish copper mines. The Coster family played an especially prominent role, acquiring shares in numerous mines and maintaining a house in Truro from which to conduct operations. They were agents to all the early eighteenth-century smelting companies at one time or another. Indeed, the Costers allowed the Bristol and Wye Valley smelters to dominate the Cornish mining industry. William Borlase, in his *Natural History of Cornwall* (1758), saw much that was positive in this business development. He credited the "Bristol gentlemen" with stimulating improvements in Cornish mining.[28] Other Cornish observers were less charitable. The mineralogist William Pryce lamented the monopoly that he believed the Costers' clients had established by the 1720s: "The four Copper Companies, viz. the Brass Wire Company, the English Copper Company, Wayn and Company and Chambers and Company, being then united and confederated, there can be no doubt of their beholding with a single eye their joint interest and pursuit."[29] There were other outside investors though. They included the Welsh Copper Company, which took over Pollard and Lane's smelting site at Neath Abbey in 1716, and had interests in a number of mines on the north Cornwall coast around Padstow, and Thomas Patten, who established a copper-smelting works near Warrington, Lancashire, in 1719, and was mining copper in the parish of Wendron by the early 1720s.[30] The result of this increased interest in Cornish copper mining was a growth in output from the 1690s onward. An estimated 81 tons of ore was raised from mines in Devon and Cornwall in 1691. By 1700, annual production stood at 1,250 tons; by 1710, 2,676 tons.[31]

Humphry Mackworth's possession of his own metal mines in mid-Wales gave him little incentive to build links with Cornwall. He began examining

Cornish copper ore supplies in earnest only after the collapse of his Company of Mine Adventurers in 1709, when he sent his Neath pay steward, Thomas Hawkins, on an exploratory visit to test copper ores raised at various Cornish mines.[32] Hawkins recorded detailed notes on numerous mines visited on this trip. Some of these were on the north coast, like the St. Merrin mine near Padstow and Mr. Eustick's at St. Just. He also headed inland and visited works at Chacewater and Redruth. At the time of Hawkins's visit, there was a sense that the Cornish mining industry was in a transitional phase; some of the mines he visited had been idle for several years and had only recently begun working again, but a number of mineral-rich mining districts had emerged, including Chacewater, North Downs, Ludgvan, Camborne, St. Just, and Gwinear.[33] This renewed mining activity was noticeable enough to attract the attention of rival European copper-mining centers, particularly Sweden, from where Henric Kalmeter was dispatched by the Board of Mines in the early 1720s on a fact-finding mission. His investigations into the mining region extended throughout Cornwall and the neighboring county of Devon. Kalmeter noted that the range of investors had spread beyond the Bristol smelters, but he recorded only one Mackworth enterprise: a smelting works in the parish of Leland near Hayle, which he leased out.[34] Despite his efforts to become better informed about Cornwall's copper resources, then, Hawkins's visit did not lead to any direct involvement in ore mining by the Neath industrialist. In this regard Mackworth lagged behind his rival smelters, who by this time were not just buying copper ores from Cornwall; they were also investing in mines there.

Mackworth may already have deduced that it was coal rather than copper ore that held the key to success of reverberatory furnace smelting on the Welsh side of the Bristol Channel. The availability of local sources of fuel meant that smelters in the Swansea District could avoid having to pay duty on coal sent by sea.[35] This tariff was first imposed in 1694 and reimposed at a higher rate in 1698. It effectively hindered the development of smelting operations in Cornwall by requiring Cornish smelters, who were obliged to bring coal in by sea, to pay double for their fuel compared to their counterparts in Neath or Redbrook.[36] Even after the removal of this financial advantage, when the tax was lifted in 1710,[37] their coalfield location continued to reap benefits for the South Wales smelters. Compared to the type of coal available to metal smelters in other parts of the country, theirs was a particularly effective fuel source for use in the metal-smelting industry. The coal reserves around Swansea and Neath, where both anthracite and bituminous seams occurred, provided a particularly good combination of fuels for copper smelting. As one later commentator explained: "Many qualities of that termed free burning or bituminous coal cannot be used alone, being what is termed too weak, while others are considered too strong; and under these two classes or designations, all the coals used in the smelting works are ranged. The two qualities of coal, strong and weak, differ considerably in price . . . but it is always of considerable importance that both qualities of coal should be used; therefore a judicious mixture of the two is a matter where œconomy in the production of copper can be practised."[38]

By the early eighteenth century, Humphry Mackworth was becoming aware that these properties were delivering considerable cost advantages over his rivals at Redbrook, who obtained their coal supplies from the nearby Forest of Dean coalfield. He found that "at Neath the coale binds hard or is very strong and durable . . . it appears that at a medium 4 tun of that coale is rather more than sufficient to work up each tun of ore. . . . But at Redbrook the coale does not bind at all and burns out very fast so that doubtless much more of that must be consumed than of the other in working up an equal quantity of the same sort of oare."[39] According to Mackworth's calculations the Redbrook works used one ton more of coal per ton of ore at a cost of 5s 6d per ton compared to the 4s per ton paid for coal at Neath. He concluded that the relative costs of smelting six tons of copper per week were £510 at Neath and £548 at Redbrook.[40]

With Neath's cost-effectiveness established as a location for coal-fired smelting, expansion and adaptation was possible. By the second decade of the eighteenth century there were two significant developments farther west, on the banks of the River Tawe, north of the town of Swansea. The first and most significant was the opening, in 1717, of the Llangyfelach works; the second was the establishment of the neighboring Cambrian works just a few years later.

Beginnings in Swansea

The Llangyfelach enterprise was headed up by Dr. John Lane and John Pollard, who relocated to Swansea after the demise of their operations at Neath Abbey the previous year. Geographically it was a minor move of approximately six miles, but it marked a new beginning. It was the first copper works to be established in what was to become the heartland of Swansea Copper. The reasons for Lane and Pollard's decision to quit Neath for an untried spot in a neighboring river valley are difficult to pinpoint. They may have been attracted in part by the opportunity to distance themselves from Mackworth's influence in Neath, but probably more important were Swansea's "pull factors." A more sizable town of around twenty-five hundred inhabitants, Swansea enjoyed easier access to the sea via its more spacious harbor. The novelist and pamphleteer Daniel Defoe, who visited both Swansea and Neath in the 1720s, was more impressed by Swansea, concluding that it was "the chief sea port" and "a very considerable town for trade." Its harbor, he noted, "sometimes sees a hundred sail of ships at a time loading coals here."[41] Around the town of Swansea coal deposits were abundant. Local landowners like Thomas Price, Walter Hughes, and Thomas Popkin opened collieries on the western side of the River Tawe in the first decade of the eighteenth century and, further west, mining was well established around Loughor, Burry Port, and Llanelli, where the coal was highly prized for its hardness and easy lighting qualities.[42] The town enjoyed greater freedom from gentry influence than other Glamorgan boroughs, thanks in part to the fact that its principal landowner, the Duke of Beaufort, was only an occasional visitor. His affairs were run by local agents who adopted a favorable attitude toward industrial entrepreneurship on the Duke's lands. Lane and his partners may also have been encouraged by the close ties between Swansea's urban elites and the commercial classes of Bristol, which served as the financial and trading center for Glamorgan's smaller towns in the early eighteenth century.[43] For the erstwhile Bristol surgeon and his west country associates, then, there was much to recommend Swansea as the location for a new commercial undertaking.

Whatever the precise motivation for the move, theirs was a serious new industrial venture. The new premises were located at Landore, some two miles upriver from Swansea harbor on a site leased from local landowner, Thomas Popkin. Often referred to by contemporaries as the "Llangyfelach" works (figure 2.2), the name taken from the sprawling parish in which it was situated, the new works was equipped with the latest water-powered technology and designed to operate on a larger scale than any of the preexisting smelting houses in Bristol.[44] By 1727 it had a labor force of forty men and two boys. A prevalence of Welsh names, such as Rees Morgan Harry, Evan Jenkin, and John Williams, among this workforce, suggests that they were either local Swansea men or had been brought to the Llangyfelach site from Neath by Lane. It was equipped with twenty smelting furnaces along with refining houses, a forge, a laboratory, and a mill for making copper rod, and cost some £2,000 to construct.[45] The potential of the new location was evident to other

Figure 2.2. South prospect of Llangyfelach Copper Works, by Robert Jacob Hamerton.
City and County of Swansea: Swansea Museum Collection

investors too. Within four years a second smelter, the Cambrian copperworks, appeared in Swansea, located to the south of the Llangyfelach premises, on property belonging to Swansea's principal landowner, the Duke of Beaufort. Significantly, the partners behind this new development were two of the Duke's local agents: his steward, Gabriel Powell Sr., and his rent collector, Silvanus Bevan.[46] The latter, a prominent Swansea Quaker, and part of a visible Quaker presence in the business communities of Swansea and Neath, and in the Bristol brass industry in the early eighteenth century,[47] may well have been responsible for engaging the services of fellow Quaker James Griffiths to run the new Cambrian copperworks on their behalf.

Despite the favorable position that the two new copper smelters enjoyed, just a short distance upriver from Swansea's large harbor, where ores could be brought in by sea and transported easily to the furnaces, it was not plain sailing. The early years of the two new Swansea enterprises were overshadowed by the financial crisis of 1720, which saw the collapse of share prices in the South Sea Company, the East India Company, and the Royal African Company.[48] At Llangyfelach, John Lane soon found himself in financial difficulties and, perhaps unable to recover the capital outlay for his new works, he was bankrupt by the mid-1720s, leaving his business under the management of his former assistant, Robert Morris. Although local Swansea historians have lauded Lane as the "cofounder" of the copper smelting industry at Swansea,[49] it was Morris who, over the next few years, did more to secure the operation of the Llangyfelach works and to establish the location as a profitable one for copper production, capable of attracting further investors. Like Humphry Mackworth, Morris was from the English county of Shropshire, close to the border with mid-Wales, and he had a Welsh wife, Margaret Jenkins of Machynlleth.[50] Having witnessed Lane's bankruptcy at close hand, he felt far from secure in his new position of responsibility. As early as 1727 he faced a fresh crisis when an outbreak of fever afflicted William Bevan and six of the coppermen, resulting in the death of one of them, Jenkin Johns.[51] Such setbacks were part of the highly insecure environment in which the early entrepreneurs in the Swansea District operated. The neighboring Cambrian works apparently faced similar disruption. Morris commented in 1727 that "copper works without smoke is a melancholy sight. It gives me anxious reflexion to see our neighbour (Griffiths) in that condition."[52] There was not yet a clear sense that copper smelting had a secure future in the Lower Swansea Valley, or that more than one copper-smelting business was sustainable there, but Morris managed to survive early setbacks. The exact fate of James Griffiths at the Cambrian works is unknown, but by 1735 Robert Morris had taken over the tenancy and the premises was later converted into a pottery manufactory.[53]

Building a Business: Robert Morris at Llangyfelach

At least part of the explanation for Robert Morris's success in riding out the early crises for Swansea Copper seems to have been his willingness to immerse himself in the enterprise and to extend a level of personal control over all aspects of the business. One example of this was his oversight of employment structures and wages. The services of key furnacemen and refiners were highly prized, and Morris sought to secure them by drawing up strict agreements with his workers. Individuals trained in the latest smelting methods were placed on contracts of three or four years' duration and sometimes required to give twelve months' notice of intent to leave, effectively binding them indefinitely.[54] Robert Morris paid Bevan a salary of £80 per year, added to which the value of the house, coal, and candles he was provided with brought his real annual earnings to around £100.[55] In contrast, most workmen were paid a weekly wage, with Morris's refiners and under-refiners earning twelve shillings and nine shillings per week respectively.[56]

He also took an interventionist approach to the sourcing of fuel supplies. Under the terms of the lease of the Llangyfelach works, Morris was obliged to purchase coal from his landlord, Thomas Popkin, who owned mines nearby. The arrangement had apparently worked well while the copper works was operated by Morris's predecessor, Dr. John Lane, but in the late 1720s, under Morris's management, smelting activity was stepped up and more coal was required. Over a three-month period in the autumn of 1728, some eighteen weys of coal per week were being consumed at the works.[57] Popkin could supply less than one quarter of this from his own mines. In common with other colliery proprietors in the region he also sold his coal to customers in the West Country, Ireland, and France, and was suspected by Morris of reserving the poorest quality residue for the smelting works.[58] The result of this situation was that, from the 1730s onward, he began to seek more control of coal supplies by taking shares in his own collieries. In 1730, Robert Morris joined with Walter Hughes and his trusted Bristol associate John Padmore to mine at Llanrhidian, on the Gower Peninsula to the west of Swansea. Such ventures not only provided a more reliable source of fuel for use in the copperworks; they could also generate excess for sale.

As well as venturing into colliery proprietorship, Morris was also the first really significant player in the Cornish copper mining industry on behalf of the Swansea District. On taking over management of the works, he quickly demonstrated a bold approach to business dealings by taking on the Bristol monopoly led by the Costers. Realizing the importance of securing access to plentiful supplies of ore, and also appreciating the need to forge good relations with key landowners in order to achieve that access, Morris went in person to Cornwall in June 1727. While there, he negotiated with Lady Pendarvis at Roskear, and succeeded in purchasing nine hundred tons of unsold ore from her mine. He followed this up with further purchases which, according to the eighteenth-century writer, William Pryce, amounted to three thousand tons in total and yielded him a 40 percent profit.[59] The ore-purchasing activities of the newcomer Morris attracted adverse attention from rival purchasers. The hitherto dominant buyer, John Coster, tried the twin approach of outbidding him for ore and forcing down the selling price of copper to harm his profits, but neither strategy succeeded in halting Morris's progress.

Robert Morris's successful negotiations in Cornwall in 1727 had far-reaching repercussions, not only for the Llangyfelach works, but for the British copper industry. Morris's son, writing an account of the period some fifty years later, said that it "was the groundwork for all of their future profits."[60] John Morton noted that it was as a direct result of Morris's ore-buying visit that a new system was introduced to govern Cornish ore sales. Known as "ticketings," this system required the smelters to submit their offers for lots of ore via sealed bids, with the sale going to the highest bidder. Few of the purchasers made their bids in person. Instead, they increasingly opted to employ agents based in Cornwall to attend the sales for them. On the face of it, this was a decisive shift away from the cozy arrangement previously enjoyed by the Bristol smelting firms who were effectively able to cooperate to dictate prices, although at least one historian has questioned whether the introduction of the ticketing system really did promote more open competition for ores.[61] Arguably more important than the overhaul of ore purchasing practices, however, was the fact that the Morris intervention was the first real sign of a shift in the balance of power in the British copper smelting industry, away from the Coster-dominated smelting works in the Wye Valley, toward a new center of dominance in Swansea. The failure of Coster's attempts to force Robert Morris out of business served to underline the greater cost-effectiveness of Swansea as a smelting location. Morris could afford to pay more for ore, or to bear fluctuations in the selling price of copper, because his production costs at Swansea were low, thanks to the abundance of high-quality local coal, and also lower freight costs than his competitors.

Despite his success in gaining a foothold in the Cornish ore market, Morris did not rely solely on this source for his crucial raw material. Alongside the Cornish ore, Morris also had access to a number of other supplies. Irish ore was being imported into Bristol in the early eighteenth century from a mine in County Wicklow which, Robert Morris speculated, was expected to raise some 150 tons per month.[62] Although this seemed a promising prospect, Ireland did not become a very significant supplier of copper ore to British smelters until the nineteenth century, when new discoveries of ore, especially at Allihies in County Cork, attracted interest from British investors.[63] New York ore was also used at Llangyfelach. Small-scale copper mining in early eighteenth-century America produced ore for export, because smelting was prohibited in the colonies. Both Morris and Mackworth bought from this source, with Morris reckoning that a profit of some 20 percent could be made from smelting it.[64] As well as these ores, supplies of unwrought copper were also available to the British smelters in the early eighteenth century. Small quantities of "Swedes and Noorts copper" featured in the Llangyfelach accounts in the late 1720s and early 1730s, but its purpose seems to have been mainly experimental. Morris was aware that Swedish copper was favored in the Dutch market, and he tried to produce copper of his own, using New York ore, that resembled it: "We will attempt the same shape and likeness without which I am persuaded we shall make no figure in Holland."[65] These trials were relatively unsuccessful, but they reveal the lengths to which Morris was prepared to go in his attempts to break into key markets.

More important than the Irish, New York, or Scandinavian copper in supplementing Cornish ore supplies in the early eighteenth century was "Barbary copper," an unwrought product from North Africa, brought in to be refined and processed.[66] The use of Barbary copper played a crucial part in the healthy output figures achieved by the Llangyfelach works in the first years of its operation under Morris's management. The works' copper accounts for the late 1720s and early 1730s show that Barbary copper accounted for a substantial share of the copper produced in the works in this period. In 1732, for example, 100 tons of the 265 tons produced was from the Barbary copper received at the works during that year. In 1733 Morris's Barbary copper purchases yielded over 162 tons of his total production of 375 tons that year.[67] These kinds of output figures could not have been achieved from the ore output of the Cornish mines alone.[68] Such was the importance of Barbary copper to the British smelters by the third decade of the eighteenth century that a number of London merchants, including Morris's chief investor, Richard Lockwood, petitioned the government for a lifting of the duty imposed on its import.[69]

Lockwood was one of three investors brought in by Morris to provide capital for the Llangyfelach copper works when he took over its running in 1727. The other two were Edward Gibbon of Putney, and Robert Corker, a Cornish MP. Initially Lockwood and Gibbon each held three-eighth shares and Corker two-eighths, but in 1728 the latter sold his share to Lockwood, making him the largest single investor.[70] Lockwood's involvement in the Barbary copper petition was just one example of the multifaceted role he played in the early years of the Llangyfelach operation. Far from making a solely financial contribution, he brought with him considerable overseas trading knowledge, as well as high-level political and commercial contacts and influence. He was one of a number of experienced overseas traders who typified the new investors in the copper-smelting ventures established in Britain from the 1690s onward.[71] A Turkey merchant, enriched by the lucrative trade with the Levant, and a member of Parliament in the Tory interest with a host of influential political contacts,[72] he has been described in one study of the financial history of London in the eighteenth century as among the most "prominent figures in the city."[73] His investment in the Llangyfelach works helped secure financial resources required to fund the first significant expansion of the business since Morris's takeover, with the erection of a new water-powered battery mill on a site upriver from the smelting house, at Upper Forest. Production of copper battery ware in the form of bars, plates, bowls, and other hammered goods was already a staple of the Bristol copper firms and it was to Bristol that Morris turned for expertise in the mill venture at Llangyfelach. In 1728 he brought the Bristol engineer, John Padmore, to Swansea to help him view potential sites for a water-powered mill on the River Tawe, and to advise on costs. Padmore had considerable expertise in the harnessing of water power and was possibly involved in the building of water-powered mills at the copperworks in Crew's Hole and Conham.[74] At Swansea he advised Morris on the best site for construction of a weir, necessary to secure a reliable supply of water to drive the mill machinery.[75] The lease of the necessary land was secured from the Duke of Beaufort and the mill was constructed and operational by 1731.

The new mill facility meant that Morris and Lockwood were ideally placed to cash in on the opportunity to export copper goods to Asia. This new market opened up in the early eighteenth century thanks to the East India Company export trade in copper, which began in 1729.[76] Prior to this, the Dutch East India Company had dominated the export trade, supplying South Asia with Japanese copper, but as Japanese domestic demand increased, their export levels declined and British copper filled the gap.[77] With its stock of ore secured from the Cornish mines by Robert Morris, the works at Llangyfelach began supplying Lockwood's London company with copper plates. The works' copper account for October 1727 to September 1728 showed a stock of over 41 tons of "fine copper in plates consigned to Richard Lockwood Esquire and Co. in London." Thereafter, the quantities increased dramatically: a consignment of over 227 tons featured in the copper account a year later, and in the year ending December 31, 1730, over 265 tons "in plates and bowles to R. Lockwood and Co." left the Llangyfelach works.[78] Not all of this was destined for export to Asia, but Lockwood and Gibbon continued to be among the largest suppliers of East India Company copper in the early years of this new trade.[79]

The opening up of the East India trade in the 1730s radically changed the business landscape for British copper smelters. During the industry's formative years, when it had been centered upon Bristol, the most important source of external demand had been in the Atlantic. New works responded to the growing demand for copper processing vessels in the Caribbean sugar sector, and to the demand for cuprous goods (manillas and Guinea rods) on African slave marts. That demand continued and deepened over the course of the eighteenth century. Had the African and Caribbean markets been the only major destinations for British copper, however, Swansea firms might have found it difficult to compete with more established works in the Bristol region that were closely integrated with the city's Atlantic trade. That Asian demand afforded new works in the Swansea District the opportunity to grow is suggested in the early accounts of the Llangyfelach partnership. These indicate that, in tonnage terms at least, Bristol sales were far less significant than those in London. Robert Morris shipped just 22 tons to his principal Bristol buyers in 1730; R. Lockwood and Co. of London, on the other hand, were supplied with 265 tons.[80] It was the East India export trade in copper that gave Morris and Lockwood a reliable customer base beyond Bristol.

The Lure of the Lower Swansea Valley

Such were Morris's early successes at Llangyfelach that his rival smelters were prompted to think afresh about the location of their businesses. The first real sign of movement came in 1731 when Thomas Coster and Co. made the first of two decisive steps toward relocating their copper-smelting interests to the Swansea District. First, Coster took over the Melincryddan works in Neath. Humphry Mackworth had died in 1727, leaving his son, Herbert, with significant debt. By relinquishing Melincryddan he could concentrate his business efforts on his substantial coal-mining interests in the region. Coster smelted copper at Neath until 1742, but his involvement in Melincryddan proved to be little more than a stepping-stone toward gaining a more permanent foothold much closer to his rival Morris's territory in Swansea. In 1736, Thomas Coster and the Bristol merchants, Joseph and Samuel Percival, signed a lease with landowner Bussy Mansel for a parcel of land at White Rock in Swansea. At Melincryddan, Coster had incurred carriage charges of 12d per ton to convey ore from the river to the hillside smelting site. The elimination of this cost at White Rock was, according to one contemporary source, "the principal reason" for his relocation there.[81] The lease granted the Coster partnership an extensive new site

> situated upon or near the River Tawe in the parish of Llansamlet in the County of Glamorgan, and all that old decayed or ruinous water or grist mill called Knapcoch Mill situated near White Rock aforesaid, together with free

> liberty and authority to erect and build houses and buildings thereon hereinafter mentioned and free liberty to pull down and destroy the said mill and to erect and build any mill or mills, engine or engines or other devices whatsoever if wanted or necessary and also the free use of the dock and quay at White Rock aforesaid for landing and laying down ore, clay or other goods and shipping of copper or other metals or goods.[82]

The lessees were committed to build, within three years, a new smelting house with at least twenty furnaces.[83] Coal was to be supplied from Mansel's own collieries at a rate of twenty-one shillings per wey and, perhaps in the knowledge that Welsh-based smelters suspected colliery proprietors of selling their best coal elsewhere, Coster and the Percivals built in a measure of quality control by stating that it should be subject to the refiner's approval "without being picked or culled of the big coal."[84] It was a sure sign that, for the new arrivals, coal was the key to success in their new venture.

It was a generous lease of a very promising site that gave the new proprietors scope to develop both smelting and processing facilities but, revealingly, it was only the coal-intensive smelting part of the operation that they chose to develop there. For processing purposes, the Costers retained their established network of mills in the Bristol area. In the year ending June 30, 1750, over 801 tons of copper was battered at Bye Mill and a further 628 tons at Publow Mill. At Publow, 647 tons of copper bar was rolled in the same year and a further 1,312 tons of copper was rolled at Swinford Mill. At White Rock itself, a long, narrow hall, known as "The Great Workhouse," was constructed to house the smelting furnaces, and a calcining house was later added.[85] By the middle of the century, it had become a major new force in the copper-smelting business. Over 360 tons was produced there in the year ending June 1751, from the smelting of over 2,747 tons of ore with 1,529 weys of coal.[86] It was a figure that compared very favorably with the Lockwood-Morris copper output of 361 tons for the calendar year of 1750.[87]

With two firms now operating on a sizable scale, on opposite banks of the river, the town of Swansea, just a short distance downstream, began to feel the effects. Population growth gathered pace, exceeding two thousand for the first time in the 1720s, thanks to in-migration as well as natural increase.[88] There were around sixteen collieries in operation, helping stimulate use of the harbor and the general enrichment of the town.[89] These developments should not be overstated: the town had poor roads and only two postal deliveries per week; commercial facilities such as banks were nonexistent.[90] Nevertheless, a small but significant shift had taken place in the world of copper. In the space of sixty years, thanks to its coal assets and its coastal access to ore, and with the help of outside capital, business skill, and lucrative new markets, Swansea had moved from the periphery to the center of the British copper-smelting landscape. It was still a long way from realizing its full potential in an industry that underwent further transformations in scale, demand, and technologies over the next century, but its newfound supremacy as a smelting location, once established, could not be easily dislodged.

Swansea's Ascendancy, 1750–1830

From the middle of the eighteenth century to the end of the 1820s, Swansea Copper underwent a transformation. From a relative upstart in the British copper industry in the early eighteenth century it had become, by the early nineteenth century, the undisputed center of copper-smelting activity. The transformation was driven by growing demand for copper to manufacture a bewildering array of products. These ranged from ship sheathing, increasingly used to protect the hulls of naval and merchant fleets in the latter decades of the eighteenth century, to the manufacture of large vessels used in the brewing and distilling trades, to boilers for steam engines, and cylinders for the printing of cotton textiles. This market diversification induced new investors, many of them with links to copper ore mining or metal manufacturing, to enter the smelting business in Swansea for themselves. Gradually, both banks of the River Tawe, to the north of the town, became crowded with new copper-smelting premises, as Middle Bank, Upper Bank, Rose, Ynys, Hafod, and Morfa works joined the existing sites at Llangyfelach and White Rock to create the most concentrated location for copper smelting anywhere in the world. The impact on the town of Swansea was double-edged: by the 1820s signs of newfound wealth and status were visible in the improved fabric and facilities of the town center, while the effect of smoke pollution in the Lower Swansea Valley was becoming an issue of public debate. But the period of Swansea's ascendancy in the world copper trade was not one of linear progression. The decades of the late eighteenth and early nineteenth centuries in particular were characterized by fluctuating prices and volatile markets. Increasingly the Swansea smelters felt the effects of political conflicts and growing technical knowledge and competition from overseas producers as they came to terms with their place as suppliers of a global commodity in the world marketplace. Ultimately their survival in the trade demanded vigilance, flexibility, and technological skill in order to keep up with the quality demands of customers and the rival products of competitors at home and abroad.

Investment and Expansion

The first real stimulus to the years of ascendancy for Swansea Copper was the East India Company's commencement of an export trade in manufactured copper in 1751. Although the company's economic impact in general has been questioned, its ability to stimulate growth in particular commodity sectors, and specific regions, such as the copper trade in South Wales, has not gone unnoticed.[1] Having provided a channel for the sale of unwrought copper in Asia since 1729, the extension of its trade to manufactured copper goods was an exciting development. Asia became established as the most important destination for copper exported from Britain in the century after 1750. East India Company copper shipments amounted to between one-sixth and one-quarter of Britain's national output by the 1780s.[2] The impact on smelting activity in Swansea was almost immediate (table 3.1). In the 1750s the original Swansea smelters were joined by two new establishments: Middle Bank opened in 1755 and Upper Bank two years later, both on the east bank of the River Tawe, north of White Rock works. The investors were Chauncey Townsend, a London merchant, and his son-in-law, John Smith, a solicitor to the East India Company, who must have had a good knowledge of the company's capacity to sell copper products in Asia.[3] Townsend was an ambitious and creative entrepreneur. The combined size of the furnace halls at Middle Bank was ten times greater than the original Swansea smelting site at Llangyfelach. His business approach was reminiscent of Humphry Mackworth's, as he attempted to integrate the sourcing of raw materials with

TABLE 3.1
Copper-smelting companies located on the River Tawe (Swansea) by the end of the 1750s

Name of Company	Name of Copper Works
Lockwood, Morris & Co.	Upper Forest
Joseph Percival & Copper Co.	White Rock
Chauncey Townsend & Co.	Middle Bank
Chauncey Townsend & John Smith	Upper Bank

Source: R. O. Roberts, "The Smelting of Non-ferrous Metals," in *Glamorgan County History,* vol. 5, *Industrial Glamorgan,* ed. G. Williams and A. H. John (Cardiff: Glamorgan County History Trust Ltd., 1980), 86–91.

copper smelting and processing. As well as his new smelting works at Middle Bank, he had a copper mill at Wraysbury, Buckinghamshire, where goods were produced for the London market. He also mined coal from his own collieries at Llansamlet and Landore, built a wooden tramway to convey the fuel to the quayside, and invested in nonferrous mining sites in mid-Wales.[4]

While Townsend and Smith's investments at Middle and Upper Bank marked a significant extension of the smelting capacity of the Swansea District, the arrival of new investors and the expansion of facilities was also taking place at some of the older sites. By 1752 Lockwood, Morris & Co. had relocated their smelting operations from Llangyfelach to a new site at Upper Forest (figure 3.1), close to where their first water-powered mill had been established two decades earlier and where there was room to expand their rolling mill capacity. Figures from surviving account books show that the move to the new site allowed them to increase production (figure 3.2), with much of the produce of the new smelting works being processed at Forest Mills before being shipped to London in a variety of forms.

Expansion was also in evidence at White Rock, which was controlled by a succession of Bristol-based partnerships supplying copper to processing mills in the city's hinterland (figures 3.3 and 3.4). Joseph Percival, one of Thomas Coster's original partners at White Rock in the 1730s, operated the

Figure 3.1. Forest copper works, Clase, near Swansea, 1786 or 1800, by Philip James De Loutherbourg (1749–1812).
Tate Gallery ©Tate, London 2019

Figure 3.2. Copper produced by Lockwood, Morris & Co. (tons)
Compiled from NLW MS15103-9 B, papers relating to Llangyfelach and Forest copper works, 1744–1789

works under his own name until 1764, when it was taken over by another Bristol entrepreneur, John Freeman. It was under Freeman's stewardship, in particular, that production at White Rock began to expand significantly (table 3.2). Direct comparison of output at the different Swansea copper works is difficult because of the incomplete nature of the surviving records, but the available figures for White Rock suggest that smelting operations were being stepped up in the third quarter of the eighteenth century, just as they were at Upper Forest.

The extension of existing works and the building of new plant led to sky-rocketing output. British copper production in the 1740s fluctuated around one thousand tons per annum. In the early 1760s, it passed the two thousand-ton mark; by the end of the 1760s, more than three thousand tons was being smelted. At the outbreak of the American Revolutionary War, over four thousand tons of copper emerged from British smelting halls.[5] There can be little doubt that the Swansea District was responsible for most of this dramatic growth. Bristol remained an internationally important center of copper production in the 1740s and 1750s, but by the time of the American Revolution its copper and brass sector was in unmistakable decline. Those conjoined trades, a prominent Bristol merchant reported in 1788, had long

Figure 3.3. Large copper-smelting works at Swansea, 1786 or 1800, by Philip James De Loutherbourg (1749–1812).
White Rock copper works from the River Tawe (*above and figure 3.4*) illustrate the advantages of the site for the off-loading of copper ore brought upriver from Swansea harbor.

been "in point of importance & capital employed" mainstays of his city's industrial base. There had been until recently "two large companies engaged in the copper smelting business & one in the brass making trade." But, he lamented, one "of the copper companies has lately declined business" and the turnover of the other smelting firm was "much more limited than it was."[6] The Swansea District, by contrast, was thriving mightily. No fewer than 310 reverberatories were at work there by 1780.[7]

Although the smelting of ore was now concentrated in South Wales, the processing of copper was not. Despite the building of battery works and rolling mills in the Lower Swansea Valley, much of the copper produced in the Swansea District was sent elsewhere to be transformed into sheets, rods, or wire. A good deal was processed in the mills of the Bristol region, which continued to flourish despite the decline of copper smelting locally. Still more was handled in mills to the southwest of London, on the Thames and its tributaries. Very often, a single partnership or a set of overlapping partnerships

Figure 3.4. The Bristol Company copper works near Swansea c. 1811, by John G. Wood.
By permission of Llyfrgell Genedlaethol Cymru / The National Library of Wales

would control the entire production chain. The English Copper Company, for example, smelted Cornish ores in South Wales at Melincryddan (between the 1740s and 1760s) and at a new site in Taibach, some ten miles east of Swansea, in the parish of Margam (from the 1770s), but it also had a rolling mill on the River Wandle at Wimbledon and maintained a riverfront depot at Vauxhall.[8] George Pengree & Co., to take another example, smelted at the Middle Bank works in the 1770s and 1780s; the firm also had battery hammers at Temple Mills on the Thames, a rolling mill at Wraysbury farther downstream at the confluence of the River Colne with the Thames, and a London headquarters at Snow Hill in the City. Integration, forward and backward, allowed smelting concerns to guarantee a market for their products, and granted brass makers and mill operators a secure source of copper.

Just as the Swansea District was building capacity as a smelting center in the 1750s and 1760s, and attracting investment from companies with allied interests in metal processing and finishing, it faced two new developments in the British copper trade that further accelerated the pace of change over the next two decades. The first of these was the discovery of a rich source of

TABLE 3.2
White Rock copper smelting in selected years

	June 30, 1750–June 30, 1751	June 30, 1760–June 30, 1761	June 30, 1764–June 30, 1765	June 30, 1769–June 30, 1770	June 30, 1774–June 30, 1775
Copper ore delivered at White Rock			2,759	4,208 (3,868 from Cornwall; 340 from "sundry places")	4,755 (including 224 tons of Irish ore)
Copper ore smelted	2,747	2,858	2,801	3,692	4,629
Copper made (tons)	360	488			

Source: Account and Memorandum Book of White Rock copper works, 12171/1, Bristol Archives.

copper ore in Anglesey; the second was the opening up of a major new maritime market for copper as a protective covering for the hulls of ships.

Copper Ore Supply: Cornwall, Ireland, and Anglesey

After a decade of prospecting and surveying, in 1768 a large deposit of copper was discovered at Parys Mountain on the island of Anglesey, off the coast of North Wales. Welsh copper smelters had spent decades hoping for the discovery of a significant domestic supply of ore to lessen their dependence on the Cornish mines, but the discovery of the Anglesey deposit brought them more than they bargained for. As well as boosting ore supplies, it also propelled a powerful new entrant into the industry: Thomas Williams (1737–1802), an Anglesey solicitor. Williams began as the lawyer for one of the Parys Mountain estates underlain by the copper lode. Before long, he had taken on the running of the mine himself, forming the Parys Mine Company to do so. He then entered into partnership with the proprietor of the neighboring Parys Mountain estate and began mining there too, now in the guise of the Mona Mine Company. Under his direction, the Parys Mountain mines made a sensational, if short-lived, impact on British copper ore production. In 1773, Cornwall produced 100 percent of the copper ore raised in Britain. Fifteen years later, with Anglesey riding high, Cornwall's share of national output had shrunk to 57 percent.[9]

Thomas Williams had no wish to be subject to Swansea's copper companies, a smelting cartel of which the Cornish mine proprietors complained regularly. The Anglesey man wanted smelting capacity of his own. He acquired it at Ravenhead in Lancashire in 1779, where the Parys Mine Company's ore was smelted using locally mined coal. The Mona Mine Company's ores were smelted at another Lancashire smelting works, that of the Stanley Company. Williams also planted his standard in the Swansea District, shipping Anglesey ore from the port of Amlwch for smelting at Upper Bank and Middle Bank, the former Chauncey Townsend works, which he acquired in 1782. With this acquisition, Thomas Williams assumed a dominant position in the trade, which he consolidated by opening his own processing mills at Greenfield, near Holywell, North Wales, supplied with cake copper from his own smelting works.[10] The combination of Anglesey ore and Thomas Williams's policy of aggressive expansion represented a considerable threat to the interests of both the Cornish mine adventurers (who were unused to competition) and the established smelters of the Swansea District (who were accustomed to dividing the market among themselves). Further expansion on the part of Williams would surely glut the market, however, to no one's advantage, least of all his own. The Anglesey copper magnate therefore managed to curb his more pugnacious instincts and enter into a market-sharing arrangement with the Cornishmen. The Cornish Metal Company, established by Williams and the Cornish mine adventurers in 1785, was to buy up all the available ore in Cornwall at a guaranteed price. This ore would be smelted on a fee basis by a consortium of smelters: three in the Swansea District, one in Bristol, and one in Cornwall. The copper returned by the smelting companies would be sold by the Cornish Metal Company at a set price. Sales were restricted to a handful of warehouses. It was "an arrangement to equalize prices, limit output and establish an agreed pattern of sales."[11]

The arrangement benefited Williams rather more than it did the Cornish mine proprietors. Because the Cornish Metal Company offered a guaranteed price for ore, there was every incentive for the mine owners to overproduce, pushing the financial burden onto the Metal Company. The company broke under the pressure, throwing the Cornish mining sector into turmoil. Thousands of miners, it was reported in 1789, were "half starving on Charity for want of Work."[12] Thomas Williams, who ran his own production network and could control his own costs, was unaffected. The Copper King, as he was now known (not admiringly), was able to dictate to the rest of the copper trade.

Thomas Williams was a powerfully disruptive force, who broke with the practice of smelting Cornish ores with Welsh coal, which had otherwise shaped the industry over the previous half century. Had the Anglesey ore deposits lasted longer, Williams might have undermined, more permanently, Swansea's domination of the industry. But the Parys mountain ore reserves were not sufficient to instigate a decisive geographical shift in activity away from South Wales. By the early 1790s the Parys Mountain mines were each struggling to yield eight hundred tons of fine copper per year, while in Cornwall, mines were reopened and ore enough to yield five thousand tons of copper was raised in 1795.[13] The shape of the copper trade snapped back to its previous position.

Having ridden out the threat from the Anglesey mines in the 1770s and 1780s, Cornwall maintained its position as the key ore supplier to Britain's Swansea-based smelting firms through the first decades of the nineteenth century. The volumes sold at the ticketings—the regular gatherings of mine adventurers and smelters' agents at which ore was bought and sold—make that plain (table 3.3).

The ticketings (so called because bids for ore were submitted in writing by prospective buyers on tickets) were lively commercial gatherings. In Cornwall, on ticketing day, "a dinner almost equal to a city feast is provided at the expence of the mines, in proportion to the quantities of ores each mine has to sell; and the adventurers, with the companies' agents, assemble together. Soon after the cloth is removed, the tickets containing the different offers of the different companies are produced and registered by the agents of both buyers and sellers . . . and the highest bidders are of course the buyers."[14] The Swansea sales, which commenced in the early nineteenth century, were later to become the trading floor for an international ore market, but in their first years of operation they dealt primarily with the sale of ores from Ireland and North Wales. They were held every few weeks, but their timing was irregular, depending upon the effect of the weather on the arrival of the

TABLE 3.3

Copper ores sold in Cornwall and Swansea in selected years (tons)

	Ores sold at ticketings in Devon and Cornwall*	Ores sold at ticketings in Swansea
1750	9,400	
1760	15,800	
1784	36,600	
1800	56,000	
1816	77,300	4,953
1820	91,500	3,725
1832	139,100	15,873

Source: Adapted from figures tabulated in R. O. Roberts, "The Smelting of Non-ferrous Metals," in *Glamorgan County History,* vol. 5,: *Industrial Glamorgan,* ed. G. Williams and A. H. John (Cardiff: Glamorgan County History Trust Ltd., 1980), 54.

* Rounded totals.

ore ships.[15] Although their principal function was the sale of ores, ticketings also provided a collective forum where representatives of the copper firms met, exchanged news, and sometimes agreed on joint commercial policies. The companies' agents were entrusted not only with making good purchases but with gathering business intelligence. Thomas Brown, the agent for Williams & Grenfell at Upper Bank, regularly returned from ticketings with news of what was happening at rival companies.[16] It was also the place where Swansea Copper, as a collective, could be addressed by outside authorities, such as in July 1830 when the deputy sheriff of Glamorgan attended the ticketings to serve a warrant against the companies for the polluting effects of copper smoke.[17]

The Swansea sales allowed the copper companies to supplement their purchases from Cornish mines with material from elsewhere. Smelters could purchase Welsh ores from Anglesey and the Nantlle Valley, as well as output of Irish mines at Allihies in County Cork, the Knockmahon mines in Waterford, and the Avoca mines in County Wicklow. Irish economic historians have tended to question the significance of copper-ore mining for the nineteenth-century Irish economy, yet the availability of ores from outside Cornwall, even in relatively small quantities, was welcomed by the smelting firms, giving them a greater range of mines to purchase from.[18] Buying ore from a variety of sources, often in small parcels, was a characteristic feature of Swansea Copper. In the period from April to July 1829, for example, Owen Williams purchased 1,727 tons of ores at the Cornish ticketings on behalf of Williams & Grenfell from seventeen different Cornish mines. At the Swansea sales from June to September in the same year, he bought 538 tons from eight mines, including Allihies and Tigrony in Ireland, and Llandudno and Drws y Coed in North Wales.[19] This pattern of purchasing facilitated the practice of mixing ores in the reverberatory furnace, "experience having taught the smelter, that by a judicious intermixture of these ores, the fusion of the poorer and more refractory ores, is more easily accomplished, than if they were put into the furnace separately."[20] There was nothing mathematical or precise about this process. As one exponent, who became head of the nineteenth-century Swansea copper-smelting firm of Vivian & Sons explained: "A more accurate mixture, calculated from the chemical analysis of each parcel of ore might be preferable; but this on a large scale cannot be practised."[21] The precise combination of ores was probably never replicated from one furnace charge to the next, bringing a built-in element of experimentation to the process.

New Products and New Technologies

The rapidly increasing quantities of ores being sold at Cornish ticketings in the second half of the eighteenth century (table 3.3), along with the brief but meteoric rise of the Anglesey ores, owed much to the new demand for copper for maritime uses. Ships that plied tropical waters suffered from the relentless attack of wood-boring marine creatures. One species of mollusk, *teredo navalis*, was ruinously persistent in this respect, forcing naval vessels and merchantmen to spend lengthy periods in port having their hulls careened. These unproductive delays had a serious impact on the operational effectiveness of warships and the profit margins of commercial ship owners. A method of shielding timber hulls from the attentions of *teredo navalis* was therefore keenly sought.

Copper sheathing, which the Royal Navy began trialing in the 1760s, was one answer. It had clear advantages. Pestiferous mollusks could get no purchase on a sleek metallic surface. Hulls remained cleaner and free of holes, resulting in greater speed and superior maneuverability, and less time was spent in dock. The one difficulty with copper sheathing lay in fastening the sheets to the wooden hull. The iron bolts that were first used corroded rapidly, the victims of a galvanic reaction with the copper and salt water with which they were necessarily in contact. The sheathing came adrift with equal rapidity. So serious was the problem that the Navy Board was on the verge of abandoning the coppering of ships in the early 1780s. Using copper bolts rather than iron would have avoided the damaging chemical effervescence, but copper had none of the natural hardness of iron. A solution was engineered by Thomas Williams, who sponsored experiments in the cold-rolling of copper to produce bolts that were sufficiently robust. By 1784, the Anglesey man was turning out forty thousand hardened copper bolts weekly at his Holywell mill, using cake copper from his Ravenhead works.[22]

The benefits of copper sheathing were obvious to all, and as soon as the new technology's reliability had been demonstrated it was quickly taken up by merchantmen sailing to warm waters. Voyage times to India and China for East India Company vessels, for example, could be reduced by 25 to 30 percent; and for slave ships, the shortened journey times cut the mortality rates among their human cargo and increased the lifespan of the vessels themselves.[23] Sheathing provided a massive addition to the existing demand for copper. In the mid-1780s, the Birmingham metal manufacturer, Matthew Boulton, reckoned that one thousand tons of copper was consumed in the "Sheathing of Shipping" at British shipyards, with a further fifteen hundred tons absorbed by the royal dockyards. This was *one-quarter* of the copper then traded by British merchants. Boulton also assumed that a significant proportion of the two thousand tons of copper exported to France was consumed by the French Navy, so it may be that the naval market accounted for as much as one-third of the copper produced in Britain at that time.[24]

The outbreak of the French Revolutionary Wars in the 1790s brought a fresh stimulus. Purchases of copper sheets, bolts, and nails by the British navy averaged over one thousand tons per year in the period 1793–1798, as hurried arrangements were made to upgrade the fleet.[25] The wartime boost has led historians to identify the 1790s as a key decade of expansion and profitability for the industry, but the spike in demand coincided with a shortage of copper ore, caused by the depletion of the Anglesey deposits.[26] The price of copper rose abruptly, and the anxiety of the manufacturers who bought the metal was sufficient to prompt a government inquiry into the state of the British copper trade in 1799.[27] The loudest voices providing evidence to the government about the state of the trade spoke on behalf of the mine adventurers and the manufacturers. These two groups claimed to be the most distressed by the high copper prices, and the most in need of government protection to safeguard their business interests. It might seem unlikely that the Cornish mine adventurers were distressed by high ore prices, yet their claim was that rising costs in the mining sector made it difficult for them to cover their expenses. Long-established mines were being driven to greater depths, at correspondingly great expense. New mines were being brought into production, but as John Vivian, the Truro banker and mining agent, explained, it would be a long time before adventurers saw a return on their investments.[28] Indeed, the miners were so strapped for cash, Vivian claimed, that they had little choice but to sell at the ticketings if they were to cover their running costs; withholding ore was not an option. As for the manufacturers, they were aggrieved about the growing cost of a key input. Representatives of the Birmingham hardware sector were particularly vocal. Matthew Boulton pressed for legislation to curtail copper exports by the East India Company, thereby freeing up supplies for the use of domestic consumers like himself. (Boulton had good reason to be outspoken; he had a government contract for a new copper coinage to fulfill.) Many blamed Thomas Williams for the shortages and high prices. It was a charge that would have carried more weight a decade earlier, when the Anglesey man had indeed exercised monopoly control over the marketing of metal. By the end of the 1790s, however, Williams's influence was much diminished.

In all the evidence gathered for the 1799 inquiry, a Swansea perspective on the situation is difficult to find. The copper smelters were believed by some contemporaries to be less vulnerable to market fluctuations than either the ore producers or manufacturers. They developed a reputation for being able to protect themselves from volatile trading conditions by acting collectively, especially to influence ore prices. The close geographical concentration of most of the smelting firms in the Lower Swansea Valley served to reinforce the idea that they acted as a bloc. Allegations that they colluded in an attempt to control prices were common. Cornishman William Pryce, who published his detailed account of the mining, processing, and selling of copper in the 1780s, described the behavior of the principal ore buying companies as "pernicious and destructive" to the interests of the trade.[29] The existence of a smelters' collective known as the "Old Company," which acted to regulate the prices offered for copper ore at the ticketings for much of the eighteenth century, is widely acknowledged. Although its activities were undocumented, the healthy gap between the selling price of copper and the purchase price of copper ore is taken as evidence of its effectiveness.[30]

Whatever the economic realities, the belief that the real profits of the copper trade were being made by the smelters induced a new wave of investors from the mining and manufacturing branches of the industry to establish smelting works in the Swansea District (table 3.4). This was especially evident in the 1790s when companies clamored to take advantage of the growing wartime demand for copper. Beginning in this decade there was a flurry of new investment in smelting works in the Swansea District, drawing in finance not from London or Bristol, which were the two main sources of smelting capital for Swansea's earlier phase of development, but from Corn-

TABLE 3.4
Copper-smelting companies located on the River Tawe (Swansea) by 1823

Name of Company	Name of Copper Works
John Freeman & Co.	White Rock
Harford & Co.	Upper Forest
Williams & Grenfell	Middle Bank and Upper Bank
Fox, Williams & Co.	Rose
Birmingham Mining & Copper Co.	Birmingham (or Ynys)
British Copper Co.	Landore
Crown Copper Co.	Crown
Vivian & Sons	Hafod

Sources: New Swansea Guide (Swansea, 1823), 79–80; R. O. Roberts, "The Smelting of Non-ferrous Metals," in *Glamorgan County History*, vol. 5, *Industrial Glamorgan*, ed. G. Williams and A. H. John (Cardiff: Glamorgan County History Trust Ltd., 1980), 86–91.

wall and Birmingham. The Birmingham Mining and Copper Company was the first from the Midlands to enter the smelting business in Swansea, establishing the new Ynys, or Birmingham, copper works in 1793. They were followed in 1797 by two more Birmingham groups of investors that established the Crown Copper works on the Neath River and the Rose Copper Company in the Lower Swansea Valley.[31] The climate of high copper prices and difficulties in securing supplies of copper was their incentive for joining the increasingly crowded ranks of copper smelters in the Swansea District. As important consumers of copper to supply their numerous metal-manufacturing firms that turned out buttons, coins, and a range of brassware, Birmingham entrepreneurs viewed entry into the smelting business as a way of exerting greater control over their trade as well as securing a share in some of the profits that the smelting firms seemed to be accruing.[32] Cornish investment, meanwhile, came from established ore mining and banking interests in the county. John Vivian of Truro inaugurated his family's long association with the Swansea District in 1800, when he went into business in a smelting works at Penclawdd, on the north Gower coast, eight miles west of Swansea, before building his own premises at Hafod, on the west bank of the River Tawe in 1809. Meanwhile Pascoe Grenfell of Penzance and Owen Williams, the son of Thomas Williams, established the formidable partnership that took over the former Chauncey Townsend works at Middle Bank and Upper Bank in 1804. The Williams involvement was significant, and an acknowledgment that, now that Anglesey ores were depleted, Lancashire could no longer rival the Swansea River as a smelting location in an industry reliant on the use of metal ores from Cornwall.[33] Grenfell's background was in banking and dealing in copper and tin ores. His long association with the Williams family began when he acted as an agent for the "Copper King" but continued long after Thomas Williams's death in 1802 through the partnership at Swansea with his son.[34] The variable extent to which the business records of these firms have survived makes it difficult to assess them comparatively, but more complete documentation for Vivian & Sons, and Williams & Grenfell, enables some insights to be gained into the conduct of these two businesses in the early years of the nineteenth century.

Williams & Grenfell was acknowledged as the leading copper-smelting firm in Britain in the first decade of the nineteenth century. It regularly headed the lists of copper ore purchasers, buying large quantities of ore at ticketings in Cornwall and Swansea (table 3.5). With two large smelting works in Swansea, mills in Buckinghamshire and Flintshire, and warehousing

TABLE 3.5
Account of fine copper purchased in Cornwall and Swansea in twelve months ending June 1825

	Cornwall	Swansea	Private	Parry's Mine	Total
Williams & Grenfell	1,263	142	7	170	1,582
Vivian & Sons	1,310	126	14		1,450
Daniel, Nevill & Co.	1,165	94	58		1,317
Birmingham Mining Co.	536	144	554		1,234
Fox, Williams & Co.	1,037	50	33		1,120
Crown Co.	912	71	2		985
English Co.	715	35			750
Freeman & Co.	637	68			705
Mines Royal Co.	438				438
Shears & Sons	213				213
Total	8,226	730	668	170	9,794

Source: The Cambrian, July 16, 1825.

facilities in Liverpool, the firm was well placed to produce a range of manufactured copper goods for the home and export markets. Fortnightly reports from Upper Bank works in Swansea for the period from June to December 1829 show that just over two hundred tons of ore was being smelted there every two weeks using some fifteen furnaces and employing a workforce of forty-nine.[35] Almost double this number was employed at the company's Greenfield mills in the same period, processing the cake copper into a variety of finished goods, including nails and bolts, as well as ingots, sheets, and other products for further refining.[36] Crucially, their possession of a warehouse in Liverpool meant that goods could be produced in bulk and supplied promptly as soon as sales were secured. Williams & Grenfell was thus ideally placed to serve the needs of customers in Britain's industrial northwest.

The uses to which copper was being applied in Britain's industrializing economy from the late nineteenth century onward were diverse. In the buoyant textile industry of the northwest of England, centered on Manchester, copper cylinders were in demand for the printing of cotton cloth. Andrew Ure, in his *Dictionary of Arts and Manufactures*, described the printing machinery in use in the Lancashire cotton mills where cloth was passed between engraved copper cylinders, about thirty inches long, in a series of processes according to the number of colors being applied to the fabric.[37] The manufacture of these copper rollers was dominated by the firm of Williams & Grenfell, who distributed them from their Liverpool warehouse to customers in the textile towns of Lancashire. Reports of stock in their Liverpool warehouse in the 1820s consistently listed "Manchester Rollers" in their record of items waiting to be shipped (table 3.6).[38] It was a very successful op-

TABLE 3.6
Stock of copper in Williams & Grenfell's Liverpool warehouse, December 10, 1829

	Tons	Cwt	Qtr
Sheathing	1	2	2
Bolts	3		2
Forged and Cast Nails	7	10	
Braziers Sheets	5	7	
Raised Bottoms	2	19	
Flat Bottoms	2	2	1
Brass Wire	7		
Manchester Rollers	15	5	
Tile Copper	1	10	
Total	45	16	1

Source: Liverpool Report, December 24, 1829, BMSS 12283, BUA.
Note: The units of weight are tons (2,240 pounds); hundredweight (Cwt); and quarters (Qtr). There are twenty-eight pounds in one quarter; four quarters in one hundredweight; and twenty hundredweight in one ton.

eration which, J. H. Vivian had to admit, had secured his rivals "the monopoly of copper cylinders for printing cottons." In fact, the Liverpool copper market was described by John Vivian as being "in the possession of Williams and Grenfell, from the vicinity of their mills at Holywell."[39]

This successful model of smelting works, supplemented by manufacturing premises and warehousing, was one that the newer Swansea firm of Vivian & Sons sought to emulate. Although previously John Vivian had been involved in a smelting partnership with the Cheadle Company in Penclawdd, at the western end of the Swansea District, there was a sense that they were relative latecomers to the business when they built a new smelting premises in the Lower Swansea Valley in 1808–1809. The new site at Hafod (figure 3.5), on the western bank of the river, almost opposite Middle Bank works, began life as a large smelting hall with twenty-four furnaces, fed with coal delivered via an adjacent canal dock.[40] Initially, the smelting of good-quality cake copper preoccupied John Vivian's son, John Henry, who had charge of production at Hafod while his father worked to secure sales. But the production of unwrought cake copper alone was never going to be sufficient to sustain a profitable business. The demand for sheathing, in particular, prompted the Vivians to invest heavily in a mill facility equipped with steam-powered rolling equipment so that they could process their own copper into the thin, flat sheets and sheathing so desired by the shipbuilders and repairers. Without this, they concluded, there would be no guarantee of securing a share of this important new market.[41]

Figure 3.5. Hafod copper works, Swansea, c. 1818.
Pencil drawing with light sepia wash, by George Orleans Delamotte (c. 1788–1861). Part of a sketchbook of thirty-three images of Swansea, Neath, and Briton Ferry. West Glamorgan Archive Service

Plans for the construction of a mill on the Hafod site were quickly drawn up and the new facility was operational by 1819, turning out sheathing in a variety of dimensions to suit specific customer requirements. One contemporary writer described the process in detail: "Copper is often reduced into sheets, for the sheathing of ships and many other purposes. The Hafod works possesses a powerful rolling mill composed of four pairs of cylinders. It is moved by a steam engine whose cylinder has 40 inches diameter. . . . The cylinders for rolling copper into sheets are usually 3 feet long and 15 inches in diameter. . . . The upper roller may be approached to the under one by a screw so that the cylinders are brought closer, as the sheet is to be made thinner."[42] These adjustable rollers were put to good use. A surviving weekly mill report for 1828 revealed that sheathing was produced in a range of different weights from twelve ounces through to thirty-six ounces. It had become a major part of the output of the mills, making up some 69 tons of the total stock of 118 tons of copper in the mill in that week.[43] In addition to the new rolling mill, a nail manufactory was opened at Hafod for making metal nails and spikes as well as brass parts for engines. In 1822 J. H. Vivian also took over the lease of Forest Mills, the premises developed by the Morris and Lockwood partnership. Located about two miles upriver from the Hafod works, the water-powered facilities there gave him access to two additional pairs of rolls for making sheathing as well as a hammer for producing battery ware.[44] But, as the Williams & Grenfell model demonstrated, the use of a well-equipped mill for the production of manufactured goods was only part of the picture; the possession of warehousing facilities where stock could be readied for rapid dispatch to customers was also essential. There was warehouse space at Hafod where produce from the mill could be stored in the short term, but a facility nearer the main copper markets was also needed. By 1811 the decision had been made to appoint a London agent, Thompson & Co., who could provide warehousing facilities and build up a customer base for Vivian & Sons' copper.[45] Speed of delivery, as J. H. Vivian recorded in his memorandum book, was crucial to securing sales: "By a mill at Hafod and a warehouse in London, we might get the greatest part of all the export orders and supply sheathing to most of the ship builders. . . . No one can wait for copper from Wales and a warehouse is absolutely necessary to meet any order."[46]

The developments at Hafod showed that it was not just naval customers and wartime demand that were the mainstays of Swansea Copper by the end of the 1810s. In fact, the British Navy, which had been an important purchaser in the 1790s, opened its own metal mills at Portsmouth and Chatham in the early nineteenth century. Not only did this give them the capacity to roll their own sheathing, but their access to large volumes of scrap copper from older vessels meant that they did not have to purchase copper from the private smelting firms.[47] The importance of sales to customers in commercial shipping took on extra significance in this context. The export market for sheathing and bolts was also shifting. While wartime conditions stimulated domestic demand for these items, conflict also had the potential to disrupt exports. After the commencement of England's war with France in 1793, shipments of copper sheathing and fastenings to the United States became less reliable at the very time when there was an American shipbuilding boom. Government incentives induced US entrepreneurs, most notably Paul Revere, to learn the techniques of copper drawing and rolling, with such success that he was able to supply copper sheathing to the US Navy by 1803.[48] Back in Swansea, Vivian & Sons felt the effects. By 1813 they were lamenting the "loss of the American market which took yearly 700 tons of copper and as much more in brass."[49] By 1823 they had received information from their agent that "the sheathing rolled in America has been found so much better than the English" and was commanding higher prices.[50]

The competition from overseas was not the only challenge facing firms that had invested in copper rolling plant for the manufacture of sheathing. By the 1820s concerns were being voiced by naval and other customers about the lack of durability of copper sheathing as well as its high cost. In some cases, sheathing was becoming worn to the point of needing to be replaced in less than five years. Anxious to avoid the heavy expense of regular refits, the Navy Board turned to the Council of the Royal Society and the officers of the Royal Mint for help in finding more hard-wearing alternatives to copper sheathing.[51] Newspapers of the period were full of reports of the experiments by some of the best scientific minds of the era with different copper alloys, to find a material that was lighter and cheaper, but tougher and longer lasting than traditional copper sheathing.[52] Humphry Davy, on behalf of the Royal Society, found that the introduction of other metals such as zinc and iron could prolong the life of copper sheathing.[53] Meanwhile, on behalf of the Royal Mint, experiments conducted by Robert Mushet led to a patent for an alloy of copper with zinc and tin for a more durable sheathing material.[54] Experimentation was also taking place inside the copper firms, with Vivian & Sons turning their Forest Mills over to the rolling of sheets from copper regulus, with the idea that this might prove more hard-wearing.[55] Vivian & Sons acquired the English patent for bronze sheathing, originally developed in France. Pascoe Grenfell & Sons, meanwhile, favored the production of sheathing from yellow metal, developed by George Frederick Muntz, which could be produced more cheaply than copper and which won the favor of the Navy following a series of trials.[56] In the end, the search for effective alternatives to copper as a sheathing material were overtaken by developments in shipbuilding and the advent of iron-hulled ships, but the flurry of experimentation in the 1820s serves to illustrate how responsive and adaptable the smelters had to be to changing customer demands.

While sheathing and its associated bolts and fastenings were examples of finished products that Vivian & Sons and other Swansea copper firms sold in large quantities in the late eighteenth and early nineteenth centuries, much of their other output was in the form of semi-manufactured goods that were sold in bulk to wholesalers who supplied different sections of the copper market. Copper shot was purchased in bulk by large firms serving the brass trades. Early in their operations at Hafod, Vivian & Sons secured an important three-year contract to supply 250 tons of copper annually to the Bristol firm of Pitt & Co.,[57] mostly in the form of "bean shot." This was a granulated copper product, similar in appearance to lead shot, and used in brass manufacture, where the greater surface area of the metal enabled it to combine effectively with zinc and calamine, the other main components in brass making.[58] Although the manufacturing process was technically uncomplicated, the exacting demands of important wholesale customers like Pitt & Co. can be seen in the instructions they issued to Vivian & Sons for their order of shot: "They desire to have forty casks made immediately to hold about 5 cwt each. They must be iron bound, very strong and have a hole at the top about five inches diameter and an iron plate made to cover it. . . . The shot of course must be made of the best copper and I should be sorry to have any fault found with it."[59]

Elsewhere copper was also in demand by firms supplying vessels to the brewing industry. Beer brewing was on the increase in the late eighteenth and early nineteenth centuries, particularly in Ireland, where legislation was passed to encourage the consumption of beer over whiskey, leading to an increase in the size and number of brewing companies.[60] By 1822 Vivian & Sons had a contract to supply thirty tons of copper per month to the London-based firm of Shears and Co., which supplied copper vessels to the brewing industry.[61] Shears had their own copper mills where the vessels could be manufactured. Smaller quantities, however, were supplied to the brewers direct in the form of large "copper bottoms" with raised sides. One Irish customer named O'Connor, for example, placed an order with Vivian & Sons for large copper bottoms of around eight feet in diameter, with raised sides, along with a flat four-foot circle of copper to form the "closing dome" of the brewing pan.[62] Inventories of copper stock at Williams & Grenfell's Liverpool warehouse indicated that they were producing similar items at their North Wales mills (table 3.6).

As far as the production of copper goods for export was concerned, Asia was the dominant market, with the East India Company controlling the purchase and sale of products. The company undertook to purchase fifteen hundred tons of copper per year for the Asian market.[63] Not surprisingly, the company's sales were eagerly anticipated by the Swansea smelting firms, for whom a share in the East India Company's order accounted for a large slice of their annual output. The company held a sale each spring where offers to supply various commodities for export were considered. Those successful in the tendering processes were offered contracts by the company in July. By the early nineteenth century, the securing of an East India Company order was perhaps the major commercial target of the year for Swansea Copper firms. Successful firms were under an obligation to supply the full amount they were contracted for,[64] and given the quantities pledged to the company, firms tended to plan their other work around this commitment. John Vivian acknowledged as much when he wrote to his son, in their first full year of trading at Hafod, to advise that "now that we know the quantity that we are to deliver to the India, we can lay down a plan of the campaign" (table 3.7).[65]

Although the East India Company sale was meant to be an open competition, the reality was that "informal pre-competition agreements between directors and bidders were often struck, and . . . some associations formed by suppliers of different commodities met before the submission of bids in order to fix prices and agree a division of orders."[66] This is exactly what the copper firms did, meeting ahead of the sale to discuss the prices to be offered and the size of order to be sought by each firm. But there were limits to their cooperation. In the early 1820s, as the market for sheathing was becoming more competitive, there were rumors that Freeman & Co. were prepared to offer the East India Company a lower price than Williams & Grenfell. John Vivian wrote to his son: "I think the companies would act wisely to continue to offer together. But if W and G [Williams & Grenfell] say nay, I would prefer joining them as the strongest."[67]

The fluctuations in the quantities of copper being exported by the East India Company (figure 3.6) broadly mirrored its general patterns of trade, which saw a rapid growth in exports in the last two decades of the eighteenth century, when company directors pushed to increase the volume of copper exports in particular.[68] The type of goods being shipped included some items that were

TABLE 3.7
Plan for copper deliveries (tons) from Hafod works, August 1809–January 1810

August	September	October	November	December	January
Pitt & Co.	Pitt & Co.	Pitt & Co.	Pitt & Co.	Pitt & Co.	Pitt & Co.
17½	17½	17½	17½	17½	17½
Baldry	India Co.	India Co.	India Co.	India Co.	India Co.
15	43	32	40	40	16
Horweed (Bristol)		Purnell	For Sale	For Sale	For Sale
15		5 (bean shot)	28	28	52
Ditto (Liverpool)					
10					
Shears	Shears	Shears			
15	20	30			
O'Connor					
8					
Total 80½	Total 80½	Total 84½	Total 85½	Total 85½	Total 85½

Source: John Vivian to J. H. Vivian, August 1, 1809, Vivian A476, NLW.

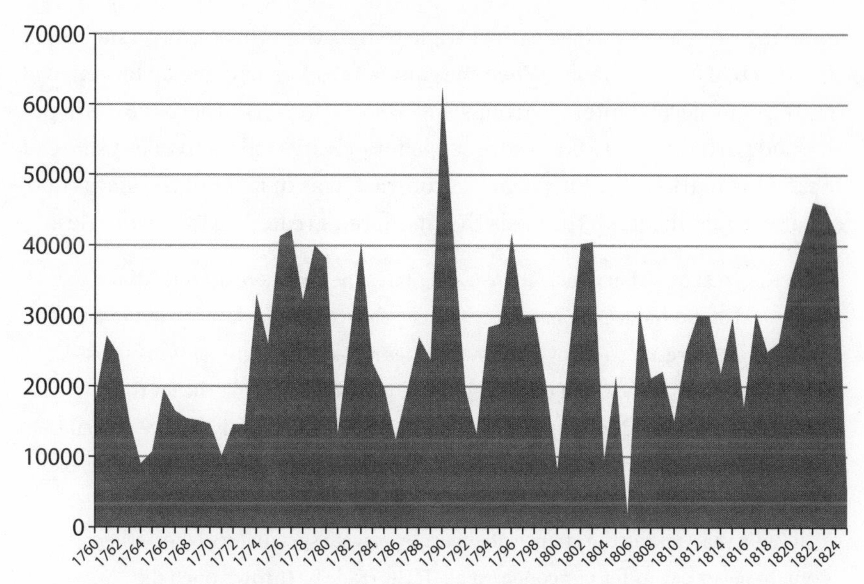

Figure 3.6. Exports of copper by the East India Company (cwt)
H. V. Bowen, *East India Company: Trade and Domestic Financial Statistics, 1755–1838* (Colchester, Essex: UK Data Archive [History Data Service], September 2007) http://dx.doi.org/10.5255/UKDA-SN-5690-1

also sold to domestic customers, but some that were manufactured specifically for the Asian market. Export figures gathered by the London Custom House in 1820 showed that unmanufactured copper, in the form of cake, made up the largest portion of copper exports to Asia, but that a range of manufactured goods, particularly sheets, nails, and wire were also being shipped.[69] Vivian & Sons' records for the early nineteenth century included notes on the prices of sheathing and copper sheets in Bombay and Calcutta, as well as references to the supply of "feathered shot,"[70] an irregular-shaped version of bean shot, which would have been destined for Asian brass manufacturers. Among the specialist products made at Hafod specifically for export by the East India Company was "Japan copper."[71] These were the ingots, six inches in length and weighing about eight ounces, that Michael Faraday watched being cast in 1819; they had a "very brilliant red colour and are sold to the Chinese as Japan copper."[72] The ability of the Swansea firms to produce items such as this, to the specified size and appearance favored in the Asian market, meant that they continued to secure orders throughout the early decades of the nineteenth century.

Not surprisingly, firms like Vivian & Sons, which secured regular contracts from the East India Company, showed less resentment toward the company's monopoly of the export trade to Asia than those whose success at the sales had been variable. When the company's charter came up for renewal in 1813, commercial interest groups and associations from across Britain petitioned parliament to allow more scope for private traders to take a share of the export market.[73] John Vivian, in contrast, was in favor of the status quo, or something like it, on the basis that it offered greater certainty of sales:

> On the great question which at present agitates the commercial world, the removal of the India Co. Charter, I was last year strongly inclined to our open trade, but there are so many great authorities and strong arguments in opposition to an open trade that I own I begin to doubt how fair it may be for the interest of the kingdom in general. For the copper trade I firmly believe it would be more safe and prudent to secure a sale for 2000 tons a year to the company or fight for as much as they might buy short of that quantity and facilities for sale in India than taking the chance of export by private speculators who perhaps might never pay us for our copper at all. If the trade be thrown open the company will buy but little from an apprehension of the competition of private adventurers and what the latter might do is very uncertain.[74]

In his willingness to put economic pragmatism before more idealistic notions of progress or the common good, John Vivian was not alone. The kinds of compromises he was willing to make over the position of the East India Company were also being made in Swansea to protect the interests of the town's growing copper-based commercial status.

The Urban Dimension

Swansea's ascendancy as a copper-smelting center took place during the classic period of Britain's industrial revolution, when mechanization was transforming the textile sector in particular, and creating new factory communities in the north of England and the Scottish central belt. Copper smelting, as witnessed in some of the products and markets it served, was linked to this process, but the nature of the society it shaped in Swansea was very different to the new textile mill towns. The upriver location of the smelting sites and their associated worker communities kept the heat, smoke, and spectacle of the growing industry at arm's length from the commercial center of Swansea. Here, some of the wealth-generating effects of the town's growing industrial success began to make themselves visible by the 1820s. Swansea remained small in comparison to the mushrooming size of English textile towns like Bradford and Preston, but population growth from six thousand to over ten thousand inhabitants in the two decades up to 1821 made it an urban center of rising significance in Wales.[75] Copper ore shipments contributed to the increased volumes of shipping using the harbor. The total tonnage of vessels entering the port of Swansea rose from just under 75,000 tons in 1790 to almost 207,000 tons in 1825.[76] Contemporaries noticed that the growing levels of trade and traffic were taking their toll on the condition of the streets, and a thorough review of paving and lighting, undertaken in 1819, resulted in a new drive to undertake street repair and introduce gas lighting to the town. The needs of the smelting firms and their employees also played an important part in the development of banking facilities, housing, and retail. Not content with the handful of Quaker-run banks operating in the town by the second decade of the nineteenth century, and unnerved by the widespread financial crisis and banking collapse in Britain in the mid-1820s, J. H. Vivian was among the Swansea businessmen who successfully lobbied for the opening of a Bank of England branch in the town in 1826. Plans for a new covered market were made in 1824 and culminated in the opening of an extensive new facility at the end of the decade.[77] The authors of the *New Swansea Guide*, published in 1823, were moved to observe that, over the two decades or so since the original *Swansea Guide* was published, the town had "astonishingly increased in wealth and population."[78]

Not everyone in the locality was so pleased with the effects of Swansea Copper. As early as the first decades of the nineteenth century it was becoming apparent to contemporaries that industrial success had an environmental cost. Travel guidebooks warned would-be summer visitors to Swansea that their enjoyment of sea bathing might be impaired by the smoke billowing from the chimneys of the smelting works located nearby.[79] More seriously, farmers whose pastures lay downwind of the smelting works began to notice an alarming deterioration in the condition of their cattle, many of whom stopped producing milk and had to feed lying down due to joint defects and other weaknesses. Convinced that it was the toxic smoke from the copper works' chimneys that was doing the damage, they eventually took their case to court in an attempt to hold the copper companies to account. They got nowhere. Indeed, it seemed that the more the copper smoke thickened over the Lower Swansea Valley, the louder were the endorsements from educated, influential quarters in the town about the benefits that the industry was bringing to the region. If air pollution was the price to be paid by Swansea in return for its status as Britain's new capital of copper, the leading industrialists and townsmen had decided emphatically that it was a price worth paying.

CHAPTER FOUR

Global Swansea, 1830–1843

At the end of the 1820s Swansea Copper underwent radical change. Hitherto, Swansea's hegemony had rested upon the exploitation of ores from the Hiberno-British archipelago itself: material from Cornwall, Anglesey, and Ireland. The use of ores from farther afield had long been ruled out by tariff barriers. The 1820s saw a series of reforms intended to liberalize Britain's previously tightly regulated commerce. Ministers took the view that Britain's industrial dominance was best served by removing export duties and clearing away obstacles to the import of industrial raw materials. William Huskisson, the president of the Board of Trade, took the lead. His tariff reforms of 1825 made a substantial cut in the duty paid on imported ore. The reduction was not in itself decisive, for duty was still charged at the "prohibitively high *ad valorem* rate of nearly 150 percent of the prevailing price of ore."[1] The really critical change came with the 1827 Customs Act, which categorized copper ore as an article that could be imported freely if the metal made from it was reexported. Swansea's copper masters leaped at the chance to "smelt in bond," processing rich foreign ores to supply their numerous overseas customers. With this new opportunity, a global field of play opened up. The gravitational pull of Swansea's furnaces was now felt everywhere.

Writing in 1848, the French metallurgist Frédéric Le Play was in no doubt that this represented an epochal change. Throughout human history, Le Play maintained, the smelting of metals had been "rigorously determined" by geology. Ores were smelted in close proximity ("almost always less than 10 kilometres") to the mines from which they were extracted, using local timber for fuel. This was a universal law to which every one of Europe's long-established copper-producing regions adhered. And yet, Le Play observed, in the "last twenty years this old order of things has changed remarkably."

The Swansea District had always depended upon imported ores, as Le Play well knew, and had therefore always been anomalous. Yet ore was brought just a short distance across the Bristol Channel—sufficiently short, Le Play must have decided, that his model still held true. After 1830 that was plainly no longer the case. All of a sudden the Welsh copper sector seemed to know "no limits other than those of the globe itself." Swansea received ores "from the island of Cuba, from Mexico, from Colombia, from Peru, from Chile, from Australia and from New Zealand."[2] Swansea Copper thus became a truly transoceanic phenomenon, involving mining and processing complexes on different continents.[3]

~

Early Victorian Swansea was not a major town in terms of population. In 1841 a little over ten thousand people lived within the old borough and the industrial suburbs that extended up the Swansea Valley. This was nothing much when compared to Manchester, or Bradford, or Sheffield. But now that furnace stuff was funneled to southwest Wales from almost every point of the compass Swansea could claim an industrial eminence that bore comparison with Manchester when it came to cotton textiles. Modest though it may have been as an urban settlement, Swansea was suddenly the center of an articulated production network that spanned hemispheres. It did more than receive shipments of ore, though. It saw the export of key technologies, most notably the steam power needed to drain mines, and the emigration of skilled technicians. In short, Swansea Copper helped to globalize the British Industrial Revolution. That was not all. Swansea Copper was at the heart of a web of credits and remittances, and it sustained a shipping network of unexampled range. Swansea therefore played an important role in advancing Britain's commercial empire in the 1830s and 1840s, especially in the southern hemisphere. This chapter will trace those developments.

Latin America was the principal source of imported ore in the 1830s. Then, in the 1840s, a series of mineral strikes in Australasia expanded the horizons of Swansea Copper still further. The opening of a mining frontier in Australia also brought about a new balance of power within what was now a global industry. Copper mining in Cuba was carried out under the direction of British capital: the companies were British-owned and headquartered in London. British influence over the Chilean mining sector was strong but not quite as stark as in Cuba. Direct investment did feature in the Norte Chico, a comparatively new area of exploitation, but in central Chile, where copper had been mined since the colonial era, funding still came from local merchants and landowners. British capital was very prominent in the mercantile sector, however. British houses in Valparaíso, for example, were responsible for assembling cargoes for Swansea-bound ore barks. Things were different in Australia. Mining was funded by an emergent colonial bourgeoisie, not by metropolitan financiers. The entrepreneurs behind wildly successful concerns like the South Australian Mining Association saw no reason to restrict themselves to mineral extraction. Rather than act as tributaries to distant Swansea they aspired to smelting capacity of their own. In 1848, Frédéric Le Play identified South Wales as the "central smelter for minerals from East and West," but even as Le Play published those words industrialists on three different continents had embarked on decentralization.[4] By the end of the 1840s smelting works that aped Swansea had been built in South Australia, in Chile's Norte Chico, and in Baltimore, Maryland.

Turning Outward: Colombia and Cuba

When the 1827 Customs Act gave Swansea's copper masters license to bring in ores from overseas they turned first to Latin America. There were good reasons for doing so. The new republics that came into being after the wars of independence were eager for foreign investment. British capitalists were equally keen to access resources and markets in the former Spanish empire; a speculative frenzy in the mid-1820s revealed just how keen. Most of the mining companies floated on the London exchange at that time had neither the technical expertise nor the local knowledge needed for success, and the boom collapsed ignominiously in 1826. The likelihood of such an outcome had been enough to deter some Swansea investors. Richard Hussey Vivian, one of the partners in the Hafod works, warned of the risks: "It seems to be walking in the dark."[5] Furthermore, the Vivians, who were closely tied to Cornish mining, were concerned that the entry of South American material would "make a strange revolution in the copper trade," the consequences of which would be harmful to their ancestral county.[6] Others were less hesitant. For all the speculative excesses of the mid-1820s, the mineral wealth of Latin America was real enough—quite enough to tempt serious investors back. Many of the more solid investments involved people already connected to the Swansea copper industry. Charles Pascoe Grenfell (1790–1867), for example, soon to be head of Pascoe Grenfell & Sons, proprietors of the Middle Bank works, had shares in silver mines in both Mexico and Gran Colombia.[7] More importantly for present purposes, Charles Pascoe Grenfell also held shares in the Bolivar Mining Association, which had a mine at Aroa, Gran Colombia (modern-day Venezuela)—a copper mine.[8] The Bolivar Mining Association began shipping ore from Aroa to Britain at the end of the 1820s. Indeed, in the early 1830s Gran Colombia was the most important overseas supplier of copper ore to Britain. Over three thousand tons of Colombian ore was landed in Britain in 1833, more than twice the contribution from Cuba. In 1835 Gran Colombia sent more than four thousand tons to Britain, an all-time high. By then, however, Cuba had caught up, and in the later 1830s, as Colombian output stagnated, Cuba powered ahead.[9]

The Spanish had known since the sixteenth century that copper was to be found in the sierra behind Santiago de Cuba, the island's second city, but early efforts to capitalize on this mineral bounty were not wholly effective. A community, El Cobre, grew up around the main excavations, but success in the seventeenth century was spasmodic, and in the eighteenth century the workings were abandoned.[10] At the end of the 1820s, however, the derelict mines became known to John Hardy Jr., a British merchant who had recently settled in Santiago de Cuba,[11] drawn there by the economic buoyancy that came in the wake of the Haitian Revolution (1791–1804) and the wars of liberation (1810–1825) in Spain's mainland colonies. The region was flush with refugee French planters and Spanish royalists who had flocked to Santiago de Cuba with whatever wealth and slaves they could extract from the territories they had fled. They snapped up estates in the city's neglected hinterland, bringing about an "increase of cultivation throughout vast tracts of virgin soil admirably adapted to the culture of the coffee plant, the sugar cane, cotton, indigo and tobacco."[12] The remnants of mining at El Cobre suggested another possibility. John Hardy Jr. "was induced, on visiting the neighbourhood for quite another purpose, to carry off some specimens of the refuse, thrown up from the old workings, in order to subject them to analysis."[13] They proved to be of extraordinary richness.

In February 1830 a partnership headed by John Hardy Jr. received a grant of the old mines from the captain-general of Cuba. The new proprietors, a mix of British and well-connected local speculators, set about restoring the mines to working order.[14] Initial progress was slow; for four years "the Mines were simply worked as a common English quarry."[15] It was not a long-term solution; if El Cobre was to be fully harnessed to the world market, more effective forms of extraction would have to be employed. But that would require investment on a scale far beyond the resources of the existing proprietors. New partners would have to be admitted—hence the rebirth in 1835 of the Cobre concern in a much expanded and more majestic form: the Company of Proprietors of the Royal Copper Mines of Cobre, with its head office at 26 Austin Friars in the heart of the City of London.[16] The modest Anglo-Cuban venture of 1830 was subsumed into a new partnership dominated by major financial and industrial personalities in Britain.[17]

The reconstituted Cobre Company represented a takeover by British capital. Or more accurately, the new company signaled the formal incorporation of the Cobre mines into the Welsh copper industry. The company's first chairman was none other than Charles Pascoe Grenfell. The board of directors included his half-brother, Riversdale William Grenfell (1807–1871), a fellow partner in Pascoe Grenfell & Sons. They were joined by Mary Glascott and her sons, who styled themselves copper merchants of Whitechapel, but who might equally as well have described themselves as proprietors of the Cambrian copper works at Llanelli, for such they were. Another Cobre director, Rees Goring Thomas (1801–1863), a banker of Lombard Street, London, also had Welsh associations. His country seat was at Gelliwernen, just outside Llanelli.[18]

Not all of those associated with the revamped Cobre Company had close connections with the Swansea District, but those without a background in the copper industry brought with them a different kind of expertise—maritime. The new copper trade of the mid-nineteenth century obeyed a global division of labor. Ore had now to be transported across vast stretches of ocean rather than the comparatively undemanding Bristol Channel. It was appropriate then that the board of the Cobre Company included George Wildes (1801–1861), the head of one of the most eminent Anglo-American merchant houses of the day, whose fleet crisscrossed the Atlantic.[19] Alongside Wildes sat John Pirie (1781–1851), a shipping magnate whose reach was genuinely global. The schooner *John Pirie* was part of the pioneer fleet of 1836 that brought settlers to the new colony of South Australia, and Pirie went on to become a director of the Peninsular and Oriental Steam Navigation Company (P&O), founded in 1837.[20]

The Cobre concern was now ready to be recast as a joint-stock colossus. It was divided into twelve thousand shares, forty-five hundred of which were to be retained by Charles Pascoe Grenfell and his fellow directors. The remaining seventy-five hundred shares were offered to the public at £40 apiece. The Company of Proprietors of the Royal Copper Mines of Cobre was therefore valued at £480,000. Investment was to sweep into the Sierra Maestra, transforming the rickety operation at El Cobre into a streamlined, technologically advanced enterprise.

This did not go unnoticed. Other investors were anxious to exploit the mineral wealth of El Cobre, and by 1836 another company had obtained a grant of mineral property adjacent to the Cobre Company's workings. The Royal Santiago Mining Company was, like its older rival, a coming together of City financiers, provincial industrialists, and shipping interests. Financial expertise was furnished by Isaac Lyon Goldsmid (1778–1859), a bullion dealer and an experienced promoter of mining ventures in Latin America; William Thompson (1793–1854), one-time Lord Mayor of London and MP for the City between 1826 and 1832; and Fletcher Wilson, a partner in the renowned banking house of Thomas Wilson & Co.[21] There were also links to southwest Wales, needless to say. Just as the Cobre Company was closely allied with the Grenfell family at Middle Bank, so the Santiago Company had its own Swansea associations, centered upon the powerful figure of Michael Williams (1785–1858).[22] Michael Williams, together with his father and brothers, was part of a consortium that took over the Rose Copper Works in 1823. Rejigged as Williams, Foster & Co., the partnership established a new smelting works in the early 1830s at Morfa, just across the Tawe from the Grenfells' Middle Bank plant. It may have been this sizable commitment at Morfa, plus the acquisition of new works in the Neath Valley in the late 1830s, that encouraged Williams to invest in Cuba as a means of securing adequate ore supplies.[23]

The Santiago proprietors, having spent an initial £35,000, followed their Cobre Company counterparts in turning their partnership into a joint-stock speculation. The Royal Santiago Mining Company was relaunched in 1838 with a share flotation. The existing proprietors were to retain one-third of the seven thousand shares issued; the remainder were offered to subscribers. In total, the capital value of the Santiago Company was set at £210,000.[24]

The Cornish Diaspora

The huge sums being raised for deployment in Cuba would be needed if the Cobre mines were to be brought up to the technological standard required for effective participation in international markets. Capital goods and skilled technicians would have to be brought in from Britain, and an unskilled labor force numbering in the hundreds would have to be mobilized locally. Cornwall would supply the specialized equipment and the skilled labor required. Indeed, the county acted as a reservoir of hard-rock expertise for the world at large in the nineteenth century, playing the role that German mining districts had fulfilled in earlier times. Both the Cobre and Santiago companies had Cornish connections to exploit. Charles Pascoe Grenfell's family was from the tin-mining parish of St. Just in Cornwall's far west, and Michael Williams of the Santiago Company had been born near Redruth in the county's central mining belt. Cornwall was very familiar territory to the key directors of both concerns.

The recruitment of miners, ore dressers, and engineers by the Cobre Company can be followed in some detail in the letter books of Alfred Jenkin, the company's agent in Cornwall.[25] Generous terms were on offer. Miners could expect £9 per month, senior pitmen could earn 10 guineas and mine carpenters could command £11 per month. Board and lodging were also provided as a matter of course. The recruits would gather at Portreath or Hayle on Cornwall's north coast to make the short crossing to Swansea and then board a copper bark bound for the Caribbean. James Whitburn, an engine man, was one of a party that sailed from Swansea on June 7, 1836; he landed at Santiago de Cuba six weeks later. He was impressed by Santiago's splendid anchorage, rather less by the fervent Catholicism of the city. "On seeing the idolotory [*sic*] of the people particularly on the Sabbath, I was constrained to remark that [Santiago] was next door to Hell."[26] (Whitburn, like many of Jenkin's chosen men, was a devout Wesleyan.)

James Whitburn's skills as an engine man were much needed. Steam engines were to take the place of the animal-powered winding gear that had hitherto drained the mines. Once sufficiently powerful engines had been installed, "the numerous hands hitherto employed in the tedious tasks of unwatering the Mines and cutting food for the cattle employed in the whims" could be "applied to the excavation and preparation of the Ore itself."[27] Back in Redruth, Cobre agent Alfred Jenkin was responsible for the export of equipment as well as miners. He ordered the Cobre Company's first beam engine from Harvey & Co. of Hayle, the master engine builders of the day, in the summer of 1836.[28] Further engines, ore stamps, and sieving machinery were to follow.

The Santiago Company followed the same course. When shares in the company were offered to the public in March 1838, the directors claimed that the "present establishment consists of one superintendant, one head and two sub-mine captains, thirty-eight miners, one head blacksmith, [and] one head carpenter, all from Cornwall."[29] Quite how many Cornishmen traveled to El Cobre in these formative years is not easily determined. This was not a once-and-for-all migration. Miners and technicians were shipped out on fixed-term contracts with the expectation of returning to Cornwall.[30] A steady replenishment of the workforce was therefore to be expected. What was not anticipated was the rate at which the Cornish workforce would be depleted by tropical disease. Late in 1838 a British guest of the Cobre Company reported that "a year before our arrival, there had been two hundred Englishmen from Cornwall; but a single sickly season had carried off the half of them, including two of the captains."[31] There was very little exaggeration in this. Yellow fever was having a devastating impact on the Cornish exiles. James Whitburn's diary, which had begun so hopefully, became a roll call of the dead. "Ben Evans died," he wrote on July 27, 1837, "one of my particular friends, and eight others all in 12 days which caused me almost to despair of ever seeing my native country and friends again." What had begun as a tropical adventure had turned into a danse macabre.

As El Cobre became a charnel house the lure of Cuba faded. "The sickness and death which have occurred at Cobre since the commencement of the present year," Alfred Jenkin admitted, "will I expect cause some shyness in the Minds of our Miners as to going there."[32] Indeed, by the early 1840s the mining companies were being forced to look beyond Cornwall. Depression-hit Wales, where so many of the mine proprietors had interests, offered an alternative. William Thompson of the Santiago Company, who was the proprietor of the Penydarren ironworks in Merthyr Tydfil as well as a City financier, took the lead. The *Merthyr Guardian* reported in March 1842 that "30 miners and 2 blacksmiths left Merthyr on the 3rd instant, for the Island of Cuba, in the employ of Mr. Alderman Thompson. With one or two exceptions they are single men, and their stipulation is for three years, or to be returned to Swansea should the climate not agree with them. They are to work in the Copper mines . . ."[33] Welsh colliers and ironstone miners were cheaper than their Cornish counterparts, but they lacked hard-rock experience, as was frequently pointed out by the mine captains who took charge of them. Employees of the Santiago Company in 1843 were derided by the local agent of the Cobre Company as "ignorant Welshmen who would be more properly described as labourers than miners."[34]

Swansea Copper, Cuban Slavery

Despite the steady influx, European workers were always a minority at El Cobre. When John Hardy Jr. provided a breakdown of the Cobre Company's workforce at the close of 1836 he put the number of "Englishmen in general, including Officers" at 80, equivalent to 12 percent of the total workforce.[35] Supporting the Cornish specialists was a motley grouping of *cobreros* (the free descendants of African slaves who had worked the mines in the seventeenth century), *isleños* from the Canaries, and Asturians from peninsular Spain. A clear majority of those who labored for the Cobre Company fell into a quite distinct category, however. Working alongside the 80 British specialists and 150-odd Cuban-Spanish free workers were 422 slaves.[36]

This reliance on slave labor was nothing out of the ordinary in nineteenth-century Cuba. The island was in the midst of a sugar boom that sucked in enslaved Africans by the tens of thousands. What to the British was an age of abolition was to the Cuban planter-class an era in which slavery was extended and intensified. Strictly speaking, the slave trade to Cuba was illegal, outlawed by an Anglo-Spanish treaty in 1817. Official disapproval did nothing to halt imports though. Sugar required slaves, so the inflow of *bozales*—as the fresh Africans were known—continued unabated, but now through clandestine channels. In truth, slavers scarcely troubled to disguise their activities. So flagrant was the contraband trade that the British pressed a new treaty on the Spanish in 1835, the year in which the Cobre Company was recast as a joint-stock leviathan. It had little effect. Nearly 108,000 *bozales* were landed in Cuba in the second half of the 1830s, most of them destined for the cane fields in the west of the island.[37] A few, however, were diverted to Santiago de Cuba for a different fate, laboring underground.

The mining companies did not advertise their use of slave labor, and for good reason: abolitionist fervor in Britain was at its height in the mid-1830s. Slavery had been abolished in Britain's own Caribbean possessions in 1834, and the 1835 treaty with Spain gave fresh force to Britain's righteous mission against the transatlantic trade. For British mining companies to be identified as slave owners would be a major embarrassment. Indeed, slaves were never to be mentioned. The Cobre Company's 1835 prospectus spoke simply of the 230 "Labourers" who were then assisting the advance guard of Cornish miners. By 1836, however, rumors about the use of slave miners had reached the Foreign Office, and the British consul in Santiago was ordered to investigate. The consul was none other than John Hardy Jr., who was faced with the delicate task of investigating his own nefarious activities. Hardy put as positive a gloss as he could on the Cobre Company's mass deployment of slaves. It was something, he claimed, that had been forced on the company: "That having found every effort to induce the Free Population to apply themselves to this branch of industry fruitless, recourse was had to the most available means of labour afforded by the Country, and purchases were then effected of Negroes distrained for debts, *irreclaimable Slaves* (who through judicious treatment have become the most steady, and placed in posts of trust), acclimated Africans, of long Standing, and all such as were dis-satisfied with their masters, and offered themselves for purchase."[38]

Hardy's claim that El Cobre offered a welcome haven for unsettled slaves convinced no one, nor did his suggestion that use was made only of long-resident ("acclimated") slaves rather than illegal *bozal* imports. A different picture of slave life emerges from the diary of James Whitburn, the Cornish engine man:

> The flogging of the Negroes in this country is most cruel. I have seen them laid on the ground, sometimes tied to a ladder, and at other times held by one man at the foot and another at the head, while another Negro with a whip 10 or 12 ft long from the end of the stick to the point of the lash, gives the Negro confined 25 blows or I may say, cuts . . . every blow rattles almost as loud as a gun. I have seen I think from 15 blows out of 25 to make cuts in the flesh from 8 to 12 inches long and open as if done with a knife.[39]

The flogging was not the end of it. The butchered victim was then fastened into stocks and left overnight "in a very painful posture . . . blood running from the cuts . . . groaning as if in a fever." In the Cobre mines such violence was mere routine. Beatings punctuated the working day.

The Cuban census of 1841 revealed that slavery was fundamental to copper mining at El Cobre. Of the 390 people employed by the Santiago Company, 56 were foreigners, 85 were free people of Cuban or Spanish origin, but 249 (or 64 percent) were slaves. The Cobre Company was significantly larger. It employed 104 foreigners, 167 free Cuban-Spaniards, and no fewer than 479 slaves (once again 64 percent of the total).[40] At that moment the Cobre Company was in all likelihood the largest slave enterprise in the western hemisphere.[41]

This scandalous state of affairs could not be disguised indefinitely. By 1841 slaveholding by the Cobre and Santiago companies had come to the attention of antislavery activists in Britain. When the British and Foreign Anti-Slavery Society launched a campaign against the continued complicity of British subjects in Atlantic slavery, special censure was reserved for the "various mining companies in this country, with large capitals at their disposal, who carry on their operations in the empire of Brazil, or in the Spanish island of Cuba . . . [and whose] mines are worked principally by slaves."[42] A parliamentary bill was introduced that would prohibit the practice. Henceforth British subjects would be barred from making use of slave labor anywhere in the world, regardless of local jurisdiction. The Cobre and Santiago companies, together with companies that mined gold in Brazil, lobbied hard against the measure and wrested two key concessions: they would not have to emancipate the slaves they already owned and, although they could make no fresh purchases, the companies would be free to hire slaves from other owners.[43] In consequence, the Act for the more effectual Suppression of the Slave Trade of 1843 was by no means the body blow that abolitionists had hoped for. The Cobre Company responded by registering ownership of its captive workforce in the name of "Don Pedro Ferrer, a Cubano and an employé of the said Company" and then leasing them back en bloc.[44] Slave labor was to remain a feature of El Cobre for as long as copper mining at the behest of Swansea's smelters continued. Swansea, ironically enough, was one of the few places in Wales where there was a significant antislavery tradition, yet nowhere in Wales did more to sustain bonded labor in the nineteenth-century world.

Deplorable though slavery was in the eyes of British officialdom, when allied to Cornish mining expertise and the thermodynamic efficiency of Cornish steam technology, it was enormously successful. Production at El Cobre shot upward in the late 1830s as the capital injected by the reestablished Cobre Company took effect and the slave workforce expanded. The company's mines produced 5,875 tons in 1837. By 1839 output was 13,655 tons, and in 1841 more than 25,000 tons of ore were raised. Output at the Santiago Company's workings followed a similar trajectory.[45]

Turning Outward: Chile

Cuban ore dominated British markets in the 1840s because the mines were directly controlled by British companies. Output could therefore be driven up quickly through the application of capital (of which the companies had a great deal), Cornish mining technology (which they had at their disposal), and enslaved labor (which Cuban conditions allowed for). Matters were arranged differently in Chile, the other major source of furnace stuff for British smelters in the 1830s and 1840s.[46] There, direct British involvement in mining was the exception rather than the norm, Cornish technology was correspondingly rare, and chattel slavery did not constitute the fundamental form of mining labor.

Mining in Cuba had been in abeyance during the late colonial period; not so in Chile. Mineral extraction was an important feature of the colonial economy. Silver was more eye-catching, but copper was a significant export commodity. Under Spanish rule, most Chilean copper was shipped to Peru and thence to other parts of Spain's American empire or to Manila, an entry point to Asian markets. Some was exported to Spain itself. After independence copper production in the new republic began to grow. The demise of colonial-era restrictions opened Chilean ports to foreign shipping, and a direct export to India, overseen by British-run merchant houses in Calcutta, started to flourish. Chilean exports at this point were of the smelted metal, not the ore. Indeed, until 1834 copper could only be exported in its smelted form. Once the shipment of ore was authorized, however, exports of the mineral rather than the metal began in earnest.[47] There was a sharp regional divide between central and northern Chile, though. The traditional smelting districts in the hinterland of Valparaíso continued to export bars; ore shipments grew but not to any great extent. The situation was very different in the provinces of Copiapó and Huasco. Firewood was hard to come by in those arid northern parts, and the rapid pace of development put a strain on such supplies as there were. With local smelters at a competitive disadvantage, shipping the ore abroad became the better option. The regional statistics are starkly revealing. In 1834, with the legal export trade only just underway, 949 tons of ore left Huasco; in 1836–1839 the annual export of ore averaged 6,246 tons. Much the same happened in Copiapó. Not a particle of ore left the province in 1834; an average of 2,560 tons was shipped annually between 1836 and 1839.[48] The upswing is mirrored in the figures for ore imports to Britain. Chilean ore appeared on the British market in the early 1830s but in parcels that were little more than experimental. Then, imports lurched upward: 1,670 tons in 1834, rising to 3,812 tons in 1835, and up again to 8,693 tons in 1836. At this point Chilean ore matched, sometimes overmatched, the Cuban on British markets.

British involvement in the mining sector itself was limited, as has been mentioned. Anglo-Chilean merchant houses in the major ports usually operated as commission agents, handling the sale of ores overseas on behalf of local mine owners. Sometimes, however, it was necessary for merchant houses to make advances to mine owners whose custom they wanted to attract or retain. From here, it was a small step for a merchant house to become a full-fledged investor (*habilitador*) in mines. That was the path followed by the house of Sewell & Patrickson, a branch of the Calcutta partnership of the same name. Sewell & Patrickson also invested on behalf of Gibbs & Co., another British-run commission agency, from the 1830s. Ores that thereby came under the control of the house of Gibbs in Valparaíso were consigned to the parent house in London, Anthony Gibbs & Sons, an important supplier of Latin American furnace stuff to Swansea at mid-century.[49]

More active intervention in Chilean mining by Britons was limited. The career of Charles Lambert (1793–1876), an Anglo-French mining engineer, is instructive in this regard. Lambert first visited Chile in the 1810s. He returned in the mid-1820s as a representative of the Chilean Mining Association, one of the new joint-stock entities that emerged from the promotional fever then sweeping the London Exchange.[50] As an alumnus of the École Polytechnique of Paris, Lambert was well equipped to introduce the latest European methods to Chile. Yet he refrained. He concluded that advancing money as a *habilitador* was preferable to implanting the capital-intensive methods used by the British companies in Cuba. In Chile, where local mining traditions were strong, there were labor-intensive alternatives.

Mineral exploitation was governed by a regulatory framework that was a legacy of the colonial era. Staking a claim was a simple matter and, provided the mine was worked fairly regularly, the miner/proprietor (*minero*) was secure in his possession. Because the barriers to entry were so low, most mines were very modest. Fixed capital was kept to a minimum. In the mid-nineteenth century "steam power had not even been dreamed of [in Chilean mining] . . . and even whims, or horse-power drawing machines, were looked upon as costly and probably unsuccessful innovations."[51] A copper mine visited by Charles Darwin in the 1830s was drained by laborers clambering up the shaft with water-filled leather pouches on their backs.[52] At another mine he watched ore carriers, shuddering with exertion, climb a zigzag of crudely notched tree trunks to reach the surface. Even so, many miners struggled to meet the barest running costs, which left them dependent upon *habilitadores*. Charles Lambert realized that insinuating his way into the traditional mining sector was a better option than importing expensive equipment, not least because the availability of alternative employment opportunities in agriculture made for seasonal fluctuations in the labor supply and the likelihood that such equipment would be underutilized.

The Central Smelter for East and West

The advent of Cuban and South American ores radically extended the mineralogical palette available to Swansea's copper masters. By 1840, the Swansea District truly was the "central smelter for minerals from East and West" apostrophized by Frédéric Le Play. Ore barks that had previously been restricted to British waters now fought their way around Cape Horn. Soon, they would race through the Roaring Forties to make port in South Australia and give Swansea's copper companies an unprecedented antipodean reach. The 1840s may then be an appropriate point at which to take stock of the industry and its workforce.

The Swansea District was now at its fullest extent. Three works lay in the far west, built on the damp flatlands that bordered the Loughor estuary: the Llanelly Copper works (established 1805) and the Cambrian Copper works (1830), both of them within sight of the open sea, and the Spitty Bank works (first opened c. 1807), further upstream. The Lower Swansea Valley had the densest concentration of smelting works. Nine were at work around 1840, all them athwart the "Swansea five-foot" coal seam. The White Rock works (1737) was the most southerly, with the Middle Bank (1755) and Upper Bank works (c. 1757) immediately to the north, along the east bank of the river. The opposite bank was also crammed with smelters: Hafod (1809), Morfa (1835), Landore (1793), the Rose works (1780), the Birmingham works (1793), and the Forest works (1752). The entire industry, except the outlying Upper and Lower Forest rolling mills, occupied a stretch of river only two miles in length as the crow flies. The Swansea Valley was also where the largest works, Morfa and Hafod, were to be found. Moving east, the Neath Valley, where the reverberatory furnace had made its Welsh debut in the 1690s, was home to two works, those of the Mines Royal Co. at Neath Abbey (1694) and those of the Crown Copper Co. (1797). The Afan Valley formed the Swansea District's eastern boundary. The Margam works, close to where the river emptied into the Bristol Channel, was long established, having been built during the American Revolutionary War. At the start of the 1840s, it was run by the Vivians. The works at Cwmavon, on the other hand, a mile or two inland where the valley narrowed, was newly opened by the upstart English Copper Co.

Contemporaries reckoned that the copper industry gave work to relatively few people. "Viewed as a field of employment of labour, the copper trade sinks into insignificance by the side of the iron trade. There are more than double the number of hands in one of the great ironworks than are employed at all the copper works in the kingdom."[53] A well-informed account of the 1860s spoke of "nearly six hundred furnaces" in the Swansea District, "employing, exclusive of colliers, about four thousand persons."[54] Hard data are hard to come by, however, and some accounts contradict themselves. In 1850, the *Morning Chronicle* described the Morfa works of Williams, Foster & Co. as the world's largest, with 1,000 employees. Yet with scarcely a pause, the *Chronicle*'s correspondent claimed that the *five* copper and spelter works operated by Williams, Foster & Co. in the Swansea District had a combined workforce of 620.[55] It hardly helped, of course, that deciding where employment began and ended in the copper industry was not straightforward. Were furnacemen alone to be counted? What of the innumerable support workers, male and female, young and old? And should the colliers employed in pits directly linked to the different works be added to the total? Opinion was divided.

On one question, however, there could be no dispute: working in copper was exceptionally demanding. By the standards of the day, early Victorian furnacemen were "well fed, well clothed and well housed," yet the debilitating nature of their work was all too clear: "Their countenances are sallow, and their persons dessicated, wiry, and thin."[56] Striking furnacemen who summarized their existence as "heat and smoke and suffering" did not exaggerate.[57] Certainly, their clothing offered little protection. Furnacemen of the 1840s wore a simple uniform: "white canvas trousers, a blue woollen shirt, and checked neckerchief."[58] It must have taken very little time for this to be drenched in sweat. Ambient temperatures within the furnace halls were high enough, but when skimming slag from the hearth or rabbling the ore furnacemen were exposed to the most intense radiant heat. They had to improvise protective clothing as best they could, perhaps by wrapping a piece of sailcloth that had been coated with wet clay around their leading arm.[59] To guard against the toxic gases that shrouded the furnace hall workers had to rely on the prophylactic power of their neckerchiefs, using them to cover the bottom half of their faces. The moment when ore was raked, red with heat, from the calcining furnaces was one of particular peril. Because the ore was not completely desulfurized, it threw off great clouds of sulfurous acid (H_2SO_3) and sulfuric acid (H_2SO_4) when brought into contact with atmospheric air. Both H_2SO_3 and H_2SO_4 are highly corrosive, irritating the nose and throat. The mist of sulfuric acid, which, being heavier than air, could not easily be dispersed, was especially damaging. It left the skin reddened and cracked; if present in high enough concentrations, sulfuric acid would even attack dental enamel. Little wonder that calciners were periodically forced to retreat to spots where they could gulp down less toxic air. "Less toxic" was the best that could be hoped for; nowhere was the air pure. By Le Play's calculation, the works of the Swansea District released 185 tons of acid into the atmosphere *every day*.[60] Arsenic was also present in "copper smoke." Volatilized during calcining, arsenic's presence was announced by its garlicky reek.[61] The "Smell ab^{t} these works," one visitor to an eighteenth-century copper works exclaimed, was "very nauseous."[62] The emission of arsenic did more than discomfort tourists, however; it put a blight on the environment for miles around. Another traveler was aghast at the environmental devastation around Neath: "The baleful effluvia blast vegetation in its infancy, and destroy the appearance of verdure in the vicinity."[63] Indeed, copper smoke was a mortal threat to the economic well-being of landowners and farmers downwind of the smelting works. Crop yields fell away; livestock sickened and died. The value of land declined in tandem. The copper companies became embroiled in a succession of legal controversies as a consequence. The Vivians were indicted for common nuisance in 1820 by the high sheriff of Glamorgan (who also happened to be a landowner in the Tawe Valley). The matter never went to trial, but subsequent indictments did, making for a series of courtroom wrangles that lasted into the late nineteenth century. The copper masters engaged the most eminent lawyers to plead their case, and hired the finest scientific minds, Michael Faraday included, to find ways of nullifying copper smoke's most damaging properties. They even found local physicians who were willing to argue that copper smoke was bracingly antiseptic; indeed, that it provided a cordon sanitaire that kept scourges like cholera at bay.[64] Judges and juries indulged the copper companies, finding consistently in their favor. Officialdom, if successive reports of the Royal Commission on Noxious Vapours in the 1860s and 1870s are to be any guide, was also sympathetic to the companies.[65]

One of the arguments put forward by the copper masters in defense of their industry was that by building stacks of stupendous height, they were able to disperse harmful smoke safely into the atmosphere. Whatever the merits of that claim, it was of little comfort to the furnacemen who worked in furnace halls suffused with toxic gases. They suffered chronic illness. In 1843, striking furnacemen claimed they lost two or three months a year to sickness, "owing to the amount of labour and the extreme heat borne, with the sulphurious nature of the work."[66] It was something the employers anticipated; at every shift change, they had substitutes on hand to step into the place of furnacemen who were not fit to work. At Hafod, no fewer than thirty-six such "helpers" were held in readiness in 1843, one for every four furnacemen. The sickliness of coppermen was proverbial, posing so acute an actuarial risk that they were "excluded from having admission to benefit societies." They felt the indignity keenly; exclusion branded them, so Hafod furnacemen complained, as "inferior and unequal to any other class of workmen."[67]

Labor Relations: "Without Apprehension of Change or Discomfort"?

Yet despite these grumbles and the arduous conditions under which furnacemen labored, industrial relations in the copper industry are often seen as temperate. At the start of the 1850s, London's *Morning Chronicle* told its readers that the copper trade was, by the turbulent standards of early industrial Britain, unusually tranquil. Employment was secure; wages were high and stable: "The effects are visible in the content and comfort of the workmen—strikes are of very rare occurrence indeed. The son succeeds the father in the works, and lives his time out, without apprehension of change or discomfort arising from adverse times."[68] This is a judgment that some historians have been happy to endorse, emboldened, no doubt, by the contrast between Swansea's apparent placidity and the insurrectionary dramas acted out in other parts of South Wales in the first half of the nineteenth century.[69] Physical-force radicalism reigned in the iron towns and colliery villages farther east, while the impoverished rural districts to the west were home to the "Rebecca Riots," described further below. Set against this rancorous backdrop, Swansea can appear calm, even torpid.

Yet there are grounds for thinking otherwise. Not least, the history of work in the Welsh copper industry has yet to be written. The Swansea District's formative years are poorly documented with respect to labor. As for the later nineteenth century, it is known that trade union organization became more firmly embedded in the Swansea area, but the involvement of copper workers awaits detailed investigation. The notion that Swansea's copper industry was a harmonious enclave in an otherwise stormy industrial world is not something that has been empirically demonstrated. It has been assumed. The basis for this assumption is the presumed stability of the trade, brought about by cooperation between companies and hereditary succession among elite workers. There is something in this. Certainly, the companies were happy to join together in trade associations, sometimes declared, sometimes covert, to put a limit on the price they would pay for ores—although price-setting arrangements were hardly unique to copper.[70] It is also true that a seniority system prevailed within the industry. Young initiates could realistically expect to graduate to higher, better-rewarded job grades as they aged. The prospect of lifelong progression, it would seem, told against workplace militancy.[71]

Seniority was an arrangement that set copper smelting apart from most trades, where apprenticeship systems were the rule. With these, a teenager underwent formal training and emerged as a fully skilled, "time-served" workman (or woman) in his (or her) early twenties. In theory, the newly qualified journeyman could go on to become a master in his own right, taking on apprentices of his own. Yet that outcome could not be guaranteed, and as the eighteenth century segued into the nineteenth, sweeping structural changes in many trades meant that journeymen found the transition to the status of master increasingly difficult to negotiate. Mechanization and/or new divisions of labor stood in the way, denying many skilled workers the craft prestige and good earnings that had accrued to earlier generations. Heightened workplace conflict was the result.

Apprenticeship was not a feature of the copper industry. Career paths took different forms and gave rise to different tensions. We are able to say so because of an exceptionally rich source for the history of work in the Swansea copper trade—the writing of Frédéric Le Play (1806–1882). Le Play was a professor at the École des Mines, the elite Parisian institution at which the most promising engineers in France received their training. Le Play was an indefatigable traveler, touring one industrial district of note after another. In 1836, he visited Swansea for the first time. Two other trips followed, leading to the publication in 1848 of *Description des procédés métallurgiques employés dans le Pays de Galles pour la fabrication du cuivre.*[72] This remarkable book, a stout little volume of nearly five hundred pages, provides an exhaustive analysis of the Welsh Process. Le Play described ten successive processes (summarized in the appendix) which he dignified with Latin numerals I to X.

What makes Le Play's book unique is its attention to the workforce. Many technical primers of the Victorian age describe the smelting of copper. They itemize the various parts of the Welsh Process; they provide the dimensions of the different reverberatories; and they specify the chemical reactions encased within. They seldom dwell upon what furnacemen did. Le Play took a different approach. He saw the human frame as an integral part of any technology. Improving a technology therefore depended upon understanding how men and women acted and interacted; hence the minute attention Le Play paid to the weights workers were required to lift, the distances they had to walk, and the inclines they had to ascend. In an age of social turmoil, Frédéric Le Play—a social scientist as well as an engineer—was also deeply concerned with order, and reconciling industrial progress with social peace was a recurrent theme of his work. The family was the key, in Le Play's view, to achieving such a reconciliation. He placed great importance on family structure and the life cycle as objects of study. Because of that focus, he is an unrivaled guide to the ways in which juveniles were inducted into the copper industry and to the roles that men—and to a certain extent women—might play as they aged.

Le Play described a seniority system in which progression began early, with puberty. Every reverberatory furnace was served by a boy of between eleven and fifteen years of age who raked ashes from the grate and wheeled clinker to nearby dumps. At the age of fifteen such boys could advance to more onerous tasks, bringing coal to the calcining furnaces (operation I of the Welsh Process) and tipping the calcined ore into storage bunkers sunk beneath the furnaces. At seventeen, these young men, now inured to heat and hard labor, were able to join in operation III of the Welsh Process, the calcination of "coarse metal." They assisted the furnacemen to collect, weigh, and load the matte produced in operation II. At nineteen, the initiates were ready to take charge of their own furnace, albeit as calciners (operation I), a role that required more muscularity than judgment, and in which they were closely overseen.[73] It was usual for calciners to continue in that role until the age of twenty-four, at which point a man was considered eligible for one of the later stages in the Welsh Process. In time, a workman of superior ability might be entrusted with operation IV, the smelting of "white metal," which was restricted to "the most moral and intelligent of the workers."[74] The most accomplished workers of all might eventually advance to the post of refiner (operation X), taking charge of the final adjustments that would reduce "blistered copper," with a metallic content of 98 percent, to a pure metal worthy of being marketed as Best Selected.[75]

Experienced refiners were well rewarded. When William Harry renewed his agreement at the Llanelly works in 1820 he was awarded £85 per annum.[76] The most trusted men might even move into management. Thomas Brown, having spent many years as a refiner at Middle Bank, became assistant to the works agent there, earning a handsome £120 salary. In 1829, Brown moved to the neighboring Upper Bank works to become managing agent in his own right.[77] Such opportunities were limited, however. By the very nature of things, only a handful of furnacemen could ever hope to become refiners. Dozens of reverberatories featured in the early stages of the Welsh Process; just a couple of refinery furnaces were required for the final process. In 1864, Middle Bank, a rather small works by nineteenth-century standards, had seven calcining furnaces, ten furnaces for smelting calcined ore, four for the calcination of coarse metal, six for the preparation of white metal, six for the roasting of white metal, and three for the roasting of blister copper. There were two refineries.[78] For most furnacemen, therefore, the exalted status of refiner was out of reach.

Wage rates reflected this hierarchy. In the industry's formative years, time wages were quite common, a device, no doubt, to retain skilled workers in the face of irregular ore shipments from Cornwall. In the 1720s, furnacemen at Crew's Hole were paid a flat six shillings a week, regardless of performance.[79] By the nineteenth century, however, piece rates were the norm. Roasters and smelters at Upper Bank were on piece rates that allowed them to take home a weekly twenty shillings at the end of the 1820s. Calciners were on a lower rate that generally yielded a little over fifteen shillings.[80] This accords with Le Play's observations for the 1840s. He thought that *ouvriers ordinaires,* aged from nineteen to fifty-five years old, could earn from twelve to twenty-five shillings per week. Workers of exceptional strength or ability commanded between twenty-five and thirty shillings.[81] Money wages were not the end of it, though; they were often supplemented by payments in kind. William Howell, who started work as a refiner at Hafod in 1811, had "two pounds of candles every week and one Wey of Coal every year."[82] Elite workers also expected a roof over their heads. Company housing was a feature of the Swansea District from the outset. "Morris Castle," a castellated apartment block built around 1770 to house employees of the Forest works of John Morris, was a remarkable experiment in industrial housing; Morriston, a grid of terraces developed a decade later, was a more conventional though still ambitious development. The forty terraced houses built by Williams & Grenfell between 1803 and 1813 for workers at Middle Bank and Upper Bank was another planned scheme.[83] Other companies improvised as best they could. At White Rock, the company subdivided the sizable house once occupied by their landlord's steward to accommodate ten different families:

> We have taken Knapcoch house near White Rock of Mr David Rees, Mr Vernon's steward . . . for the use of our Workmen and parcelled it out to the following persons:
>
> Griffith James—the parlour and Chamber
> Morgan David the kitchen
> Morgan Own the chamber over
> Cath Leyshon the little parlour
> Eliz Thomas the hall
> William Jones the passage chamber
> Thomas Phillip the little parlour
> William Rees a garrett
> Rich Thomas another Garrett
> William Owen the Brewhouse.[84]

When makeshifts of this sort failed, as they inevitably did during periods of rapid expansion in the industry, employers took to making top-up payments to workers for whom company-owned cottages were not available. Thus William Howell, the Hafod refiner mentioned above, had "ten guineas every year in lieu of house and Garden," in addition to his candles and fuel allowance. At Upper Bank, it was the custom "to allow the furnace men which have no houses of the company 6d per week towards house rent."[85] Eleven of the twenty-five furnacemen at the works fell into this category in 1830. Some furnacemen had the opportunity to take on additional work, supplementing their basic earnings. Calciners, who worked a day-on / day-off shift pattern, were especially well placed to do so. Discharging ore from barks moored in the river was a regular means of bringing in extra money during their days off. "I can work at unloading vessels," a calciner at Hafod told an investigator from the *Morning Chronicle* in about 1850, "by which I earn sometimes 4s in the course of twelve hours."[86]

Yet to focus on the wages earned by individuals can mislead. Household earnings were what determined an individual's well-being. Adult men earned the most, but many households were sustained by the wages brought in by their female and juvenile members. "The entire population," Frédéric Le Play claimed, "finds employment in Welsh copper works." Teenage boys attended every furnace, as we have seen, raking out debris and fetching coal, while women "from 20 to 40 years old" were "entrusted with the transporting of minerals and intermediate products within the works." Women were preferred to horses, Le Play explained, because they could be more precise in their movements within the narrow furnace halls, and were smart about it.[87] They were certainly not allowed to tarry. At the works visited by Le Play, a team of seven women was engaged in bringing ore to the calcining furnaces. Over the course of a ten-hour shift, they wheelbarrowed 150 tons of material between them. Le Play, whose enthusiasm for time-and-motion studies anticipated that of Frederick W. Taylor, reckoned that each *rouleuse* pushed a barrowload of 336 pounds over a distance of forty meters, moving at a speed of one meter per second.[88] Women were also responsible for removing the tons of slag tapped daily from the furnaces. Female "slag trammers" spent their days pushing tramloads of still-hot waste to the tops of ever-growing slag heaps (or, where slag heaps had already taken on mountainous proportions, to the foot of an inclined plane, up which the trams could be winched). The labor in this should not be underestimated. Copper works produced far more slag than copper. The Middle Bank works in 1828, for example, occupied a seven-acre site, but its slag dump covered twenty-six acres.[89]

The role of women in the copper industry, as in many other heavy industries, was becoming increasingly contentious in early Victorian Britain. This was, after all, the age of the 1842 Mines Act, which prohibited the employment of women underground in the coal industry. A drift toward the exclusion of women workers is evident in copper too. Le Play, who made his observations between the mid-1830s and mid-1840s, gave no sense that slag tramming by women was controversial. By the early 1850s, the mood had changed. A visitor to the Hafod works found that women's labor was being phased out. "As they marry off, or die, they are not replaced by females, but by boys. Formerly the 'slag-tramming' was performed entirely by women. I was glad to find that the system of employing females in the laborious duties they have here to perform is gradually dying out."[90]

Slag tramming, whether performed by a woman or a man, was but one of many support roles. Not everyone in the industry was a furnaceman. William Jones, managing agent at the Hafod works in the 1830s, recorded the wages paid to over twenty categories of ancillary worker: "Stock Takers, Coal weigher, Watchmen, Engine Man, ore filler . . ."[91] Some of these roles (ash wheelers, for example), were performed by boys who might be regarded as aspirant furnacemen, but most of the outlay went toward common laborers. Copper works also employed large numbers of artisans to maintain the plant: smiths and their strikers, masons, and carpenters. Taken together, these support workers were roughly equal in number to furnacemen. The Upper Bank works, a modest operation by contemporary standards, had forty-nine men and boys on the payroll in the autumn of 1829. Only twenty-five of them were furnacemen.[92]

Beyond the works themselves, there was a penumbral workforce of some size. Hundreds of colliers worked on the coal measures to the north, some at independent mines, others at collieries that were directly tied to a particular works. In the early days of the industry, it was common for copper companies to contract with local colliery owners. As we have seen, the Llangyfelach works, when Robert Morris assumed control in the 1720s, took its coal from Thomas Popkin, who also happened to own the land on which the copper works stood. Such an arrangement was not always satisfactory: landowners would often demand that the copper companies buy only from them, as Popkin did. Conversely, copper masters were apt to complain that the supply of coal was inadequate or too erratic for their needs.[93] Given the importance of a reliable supply of fuel, and the sometimes barbed relationship between the copper companies and local landowners, there was an inescapable tendency toward vertical integration. This tendency became marked in the 1830s as the industry underwent rapid growth ("all the Works being very brisk with so much building on both sides of the River").[94] Small undercapitalized collieries could not keep pace. The response of the two leading firms, the Vivians and Williams, Foster & Co., was to collaborate in the Swansea Coal Co., which acquired a number of mines in the Lower Swansea Valley, employing nearly 450 colliers by 1841. When the Vivians acquired the Margam works in 1839, they took the colliery at Goytre too as part of the bargain, together with the 90 men who worked there.[95]

The copper industry also gave employment to a large body of seamen. From the very outset, the Swansea District relied upon seaborne ore, carried on vessels that shuttled across the Bristol Channel, carrying coal to Cornwall and taking copper ore as a return cargo. An outward trade in coal was already well developed at the start of the eighteenth century, even before the copper industry took hold. Ships carrying coal cleared the port of Swansea on 515 occasions in 1701. Another 422 coal carriers cleared Neath in that same year.[96] The southwest of England was a major market. There were ports and inland towns in need of domestic fuel, and industrial activities, like salt boiling along the Somerset and Devon coasts, in need of energy. The spread of steam engines across Cornwall, which underpinned the expansion of copper mining, added greatly to demand. Sustaining this trade called for a substantial fleet. The ships themselves remained relatively small, however. In the mid-1780s, when Cornish mines were producing around thirty-five thousand tons of ore annually, Matthew Boulton took craft of just one hundred tons capacity to be standard on the Swansea-Hayle run. The crews were correspondingly small: a master, a mate, two seamen, and two boys.[97] The growth of the trade required that such vessels grew in number—there were one hundred sailing out of Swansea in the 1820s—but not greatly in size.[98] Those used in the international ore trade that sprang up in the 1830s were much larger (figure 4.1). The *Morning Chronicle* described "barques of 500 to 1000 tons burden," crewed by sixteen to twenty-five hands. Such vessels had to be well crewed, for the voyages they made were epic, tests both of endurance and seamanship: "A ship trading around Cape Horn, with Chile, makes a voyage out and home in from eight to ten months; a vessel trading with Australia makes one voyage, and a vessel trading with Cuba two voyages in a year."[99]

Figure 4.1. Cobre Wharf, Swansea, 1846, by Calvert Richard Jones.
This calotype by the pioneering photographer Calvert Richard Jones shows barks that have brought ore from Cuba grounded at low tide alongside the Cobre Wharf.
City and County of Swansea: Swansea Museum Collection

Being too large to float upriver to the copper works, barks arriving from Santiago de Cuba and Valparaíso discharged at quays along the lowest stretches of the Tawe, close to Swansea town itself. Here, another auxiliary workforce was summoned into action; its task was to dress the ore before it was taken by barge to the different works upstream. It was hard, monotonous work. At Richardson's yard, the *Morning Chronicle*'s correspondent saw men and women pounding the ores with "large flat-headed iron hammers, and passing the fragments through sieves." The workforce here, mostly Irish refugees from the Famine, were out in the open. Their counterparts at the Cobre Company's yard were a little luckier; they were sheltered under "an extensive roof, supported on iron pillars," beneath which steam-powered rollers crushed the valuable Cobre ores ("some of a brassy lustre, some dark green, and others of a rich grey").[100]

So, the furnaceman was but one in a wider cast of characters. He was the talisman, nevertheless. He presided over the central drama in copper making:

the conjuring of metal out of rock. Yet to speak of furnacemen without careful qualification is to assume that the experience of work within the smelting halls remained constant over time. It did not. Thomas Cletscher, a Swede who inspected the Conham works near Bristol at the end of the 1690s, described the nascent Welsh Process as having just three elements: roasting, melting, and refining. By the nineteenth century, however, the Welsh Process had evolved into a much more complex multi-stage operation, featuring an elaborate sequence of roastings and meltings. The Swansea copper master J. H. Vivian, writing in the 1820s, specified eight distinct steps.[101] The industrial chemist James Napier, in the 1850s, spoke of a six-stage process as conventional.[102] Le Play identified ten stages, as we have seen, but distinguished between an "ordinary process" of seven stages and the "extra process," which consisted of the full ten. This divergent testimony doubtless reflects different evolutionary paths at different works. No one works exactly matched another. John Vivian, surveying the Swansea District in the first decade of the nineteenth century, saw no uniformity. He thought the furnaces at Middle Bank unusually small and the conduct of affairs at the Neath Abbey works quite singular.[103] Two decades later, the works manager at Hafod noted how distinct the practice at his own works was from that of the Llanelly Copper Company. Yet for all the variety, it is certain that the Welsh Process, whatever evolutionary track it took, grew in scale as well as complexity. The size of reverberatories is a useful index of this growth. Thomas Cletscher reported furnace hearths that were just three feet square at the end of the seventeenth century.[104] In 1780, however, Matthew Boulton of Birmingham described reverberatory hearths that were nine feet long and five across.[105] Subsequent decades saw further development. "The ordinary form of a fusing-furnace," one nineteenth-century authority declared, "is 13 feet long by 8 wide inside measure."[106] In other words, hearth size grew by a factor of twelve between the 1680s and the 1850s (from 9 square feet to 108 square feet). This scaling up had important implications for workloads. Writing in the 1780s, Matthew Boulton noted that furnaces and the work done at them had undergone a remarkable inflation within living memory: "The old Coppers furnices would only contain 5 [hundredweight] of Ore & that Charge took 12 Hours but now they contain 7 [hundredweight] & they charge 3 times in 24 Hours."[107] To put it in a different form, furnacemen had gone from handling 0.5 tons over the course of twenty-four hours to handling 1.05 tons.

The demands made upon furnacemen were growing: they had to handle more and more material, and they had to do so with greater and greater expedition.[108] Accomplishing this required a good deal of experimentation with workplace practices. What, for example, was the optimal working day? Nineteenth-century copper works operated around the clock, lest the furnaces cool. Sunday was the only day on which work was regularly suspended. Respect for the Sabbath was not absolute, however. Operation III, the calcination of coarse metal, which took thirty-six hours to complete, had perforce to continue through the Lord's Day, and operation IX, the preparation of "blistered copper," did so too, according to Le Play, for reasons he did not divulge.[109] The standard week began at six o'clock on Monday morning and continued until four o'clock on Saturday afternoon, making for a workweek of 130 hours.[110] Most furnacemen in the 1840s worked a twelve-hour shift, changing at six in the evening and six in the morning. Historically, though, other patterns prevailed. Longer shifts may have been introduced as the industry expanded in the late eighteenth century. The manager of the Mines Royal Company noted that the men at his Neath Abbey works "began working 24 hours at a stretch" in February 1796.[111] By the 1830s, however, the manager at Hafod referred to "working the Longwatch" as the "old system." In the 1840s, Le Play spoke of the twelve-hour shift as customary, with the exception of calciners (operation I) and furnacemen at operation III, the calcination of the coarse metal. Both worked around the clock. Starting work at six in the morning, calciners spent the first five hours of their shift wheeling away calcined ore that had been pulled from the hearth several hours earlier and that had now cooled to a manageable temperature. Thereafter, periods of repose alternated with spells of extreme exertion: "Every two hours we are called by the watchman to stir the calciner, which takes a quarter of an hour. At ten at night we are called to 'pull out'; this takes one hour, and 'recharging' takes one hour; it is then twelve at night. The watchman then calls us to stir at two, at four, and at six o'clock, when twenty-four hours are up."[112] The fate of calciners was never to experience the circadian cadences of regular life.

Calciners were not just concerned with the calcining process (operation I); they were also required to assist the refiners who presided over operation X, in which very large volumes of material had to be loaded quickly into the furnace hearth. A calciner, the ever-precise Le Play calculated, spent 18.2 percent of his time charging the refining furnaces.[113] This pattern, of being summoned from one function to assist in another, is not to be wondered at. As copper production in the Swansea District grew, and as furnaces were scaled up, managers introduced new divisions of labor to cope with the bottlenecks that emerged. Once, calciners must have been able to handle the calcined ore they raked from their furnace single-handedly; by the 1840s, they could not. During daylight hours, the calciner was joined by three teenage barrow-wheelers, each of whom bore away ten tons of calcined ore during their ten-hour stint.[114] Their presence enabled the older, more robust calciner to lend a hand at the refinery when called upon. Yet the use of greater numbers of workers made furnace halls increasingly crowded. That is why nimble humans were preferred to horses for most of the carrying within the works. It is why only the brawniest of men were hired to remove the granulated matte that was the product of operation II. Employing the most strapping adults available minimized the numbers needed and allowed "the freest possible circulation through the *atelier*."[115] Indeed, operation II produced particular difficulties. The smelting furnaces were much smaller than the calcining reverberatories of operation I; they had therefore to be more numerous if they were to consume what the calciners produced. Yet the smelting furnaces were also fast acting; a charge of calcined ore could be smelted in four hours and twenty minutes. To accommodate this schedule, materials had to be hurried in and out in a carefully choreographed sequence. Le Play described how smelters at adjacent furnaces were grouped in teams of four in order to coordinate their movements. Each of the quartet went to the weigh-house in his turn, collected his next charge of ore, and returned briskly to his furnace, carrying a full hundredweight on his head.[116]

Control over the bustling furnace halls was essential for the copper companies. Theirs was a capital-intensive business, abnormally so. The companies had sunk tens of thousands into furnaces, mills, wharves, tramways, and ore yards. Thousands more were embodied in stocks of ore. This capital had to be set in motion quickly, and, if it was to return a profit, every particle of copper had to be detached from the ore into which it was bound. Close supervision of the production process was essential. Yet it was not easy to achieve. Some parts of the Welsh Process lent themselves to managerial oversight; others did not. Calciners, as we have seen, were subject to particular scrutiny. The workers did not decide when ore was to be rabbled; the watchman who supervised them did. And calciners who flopped down to doze beside their furnaces knew that they would be roused at regular intervals through the night when rabbling was to be done. They were not required to exercise judgment of their own. Things were very different with the next part of the process, the smelting of calcined ore. Here, everything was at the discretion of the furnaceman, or rather the two furnacemen who worked alternating twelve-hour shifts, as Frédéric Le Play explained:

> Each smelting furnace in operation II is served by two workers, men of between 25 and 55 years of age, each of whom works alone for 12 hours. . . . [Yet] the work of one is not independent of that of his comrade: the negligent worker who allows the furnace temperature to fall below the proper level, who permits materials to accumulate in unworkable lumps in the hearth, etc., leaves a serious problem for the following shift. Besides, it usually happens, as will be indicated later, that two shifts contribute jointly to the working up of certain charges. The conduct of work at a smelting furnace involves, therefore, a genuine association between the two workers to whom it is entrusted. The works managers therefore allow the furnacemen to choose their own partners.[117]

Indeed, smelters exercised considerable independence. There were only two points in operation II at which supervisors could intervene. One was at the weighing out of materials. "Left to themselves," Le Play claimed, "the workers, who are interested in handling as many charges as they can in the time given, will be led to neglect the most refractory ores, and to restrict the weight of the charges."[118] The other point of managerial intervention came once the charge had been smelted and the slag drawn off. Scoria from each furnace was deposited with a foreman for his inspection. Hammer in hand, the foreman cracked open each loaf of slag and pored over the fracture, paying minute attention to the grain and color of the material exposed. That allowed him to determine how much copper had eluded capture during the smelt. If the metallic residue exceeded a prescribed limit, the guilty furnaceman would be obliged to perform an additional smelt at the week's end without pay.[119] Managerial control over operation II was therefore indirect; it was confined to measuring inputs and assessing outputs. Operation III (the calcination of coarse metal), on the other hand, was subject to the "assiduous control" of supervisors, one for every four furnaces.[120] Operation IV (the making of white metal) differed again. It was a complex task. Rich foreign ores, oxides, and carbonates with a metallic content of between 20 and 30 percent were added to the calcined matte produced in operation III, together with slags collected from operations IX and X. Responsibility for assembling such a charge had to be ceded to the furnacemen themselves because, Le Play reported, "there is no means of exercising effective control over their work"; hence the necessity of appointing the "most moral and intelligent" workers to the job, rewarding them well, and allowing them free rein.[121]

Ceding control to furnacemen was not done willingly. Wherever possible, the masters aimed at close control. They levied fines on furnacemen who failed to present themselves at the start of a shift, a shilling a time at the Hafod works in the 1840s.[122] Those who left work without authorization were also subject to stiff fines. The code of conduct introduced at the Birmingham Copper Company's works in 1797, for example, threatened men on the night shift with the loss of a quarter of their wages if they quit before the ringing of the 6 o'clock bell.[123] And, as we have seen, unpaid shifts were imposed on those who did not meet performance targets. Indeed, points of conflict punctuated the Welsh Process. Knowing that they would be held to account for work deemed inadequate, workmen insisted on being issued with the proper materials for the job. Poor quality coal was a constant bone of contention. "The Furnacemen," the Mines Royal manager noted in his diary in December 1797, "on account of bad Coals, left their work this morning."[124]

Workplace frictions of this sort were commonplace in the early industrial age. Skilled workers knew their worth. They knew what wages were current across the Swansea District, and they reacted angrily if shortchanged. The furnacemen at the Mines Royal works walked out in January 1798, complaining that "they do now work for less money than Roe & Co.'s men" at neighboring Neath Abbey. They returned to work a week later, once the manager had "promised them 6d per day more."[125] As John Vivian of Hafod later remarked, "If one company is obliged to advance others must do the same."[126] Skilled workers were also strongly averse to the kind of regular attendance the masters demanded. They exasperated managers by their attachment to traditional holidays, leaving furnaces "dead" for days on end. Christmas 1837, the Hafod works agent complained, saw "men scarce & nice" (that is, noncompliant). Never had there been "such a drunken Xmas week"; there were "6 or 7 dead Furn at a time."[127] Absenteeism was also a feature of the harvest season, when bumper earnings were to be had in the fields. For that reason, copper companies contemplating wage reductions usually waited until the harvest was in before acting. In the summer of 1830, for instance, copper companies across the Swansea District were "agreed that the wages now paid are higher than are paid in any other trade, and that a reduction should be made."[128] There was danger in acting prematurely, however, as the Crown Company demonstrated when it unilaterally cut wages at its Neath works: a walkout of the workforce ensued. It would be better, the employers at large decided, if they stayed their hand. They "thought it best to leave the wages as they are now, until the Harvest is over."[129]

The masters found it easy to act in concert. They met regularly at the Swansea ticketings to bid on parcels of ore. The spirit of cooperation that ensured prices were kept within bounds was readily extended to regulating labor. In dealing with working people, the masters had considerable legal powers at their disposal. Labor legislation in England and Wales, starting with the Statute of Artificers in 1563, assumed a fundamental asymmetry between the rights of employers and the rights of those they employed. It had three guiding principles: "The first was the idea that the employment relation was a matter of *private contract* . . . between an employer who thereby acquired the right to command and an employee who undertook to obey. The second was the provision for *summary enforcement* of these private agreements by lay justices of the peace or other magistrates, largely unsupervised by the senior courts. The third was *punishment* of the uncooperative worker: not damages to remedy the breach of contract, but whipping, imprisonment, forced labor, fines, the forfeit of all wages earned."[130]

The criminalization of worker resistance was counterbalanced to some degree by the powers given to magistrates under the Statute of Artificers to regulate wages. By the late eighteenth century, however, those wage-fixing powers were falling into disuse. In 1813, they were abolished altogether. Parliament was very ready, on the other hand, to grant new coercive powers to employers. The Regulation of Servants and Apprentices Act of 1746, for example, dealt with categories of workers not explicitly mentioned in the Statute of Artificers. Like the act of 1563, it differentiated between breaches of contract committed by masters and those by servants. Masters were subject to fines; their workers faced imprisonment and hard labor. The power to incarcerate workers was one to which employers were deeply attached, and one they wanted endlessly restated. A further Regulation of Apprentices Act in 1766, which authorized magistrates to jail "any artificer, callicoe printer, handicraftsman, miner, keelman, pitman, glass man, potter, labourer, or any other person" who failed to complete their contract of employment, was little more than a reproduction of the existing act of 1746.[131] MPs were also instinctively hostile toward workers who formed unions to advance their interests. "Combinations" had long been subject to legal sanction, being considered a form of conspiracy under the common law.[132] During the eighteenth century, however, collective action by workers began to be penalized in more precise ways. Employers' groups who petitioned Parliament were rewarded with acts that outlawed combinations in specific trades and authorized magistrates to impose summary punishments. At the end of the 1790s, an inflationary and politically troubled decade, Parliament passed an all-embracing Combination Act, which outlawed *all* collective bargaining. The act of 1799,[133] somewhat refined in 1800, allowed offending workers to be jailed on conviction before a single magistrate. It was a critical addition to the legal powers already at the disposal of masters in the Swansea District.

A dispute that broke out at the Middle Bank works in 1820 gives some idea of the extent of those powers. When "coppermen" struck for higher wages on August 15, 1820, Williams & Grenfell took immediate counteraction. The company had Swansea's magistrates prosecute thirteen furnacemen under the 1800 Combination Act[134] for a conspiracy to raise wages; the thirteen were duly convicted at the Glamorgan quarter sessions on August 21. Two of them were sentenced to three months in Cardiff jail; the remaining eleven were given two months' hard labor at the Cowbridge house of correction.[135] They were not behind bars for long. They were "rescued from Gaol by their fellow Workmen" within the week.[136] The jailbreak spread alarm among the copper masters. "I am very sorry to hear of Riotous Proceedings among the Middle Banks Men," wrote John Vivian, owner of the Hafod works, "For such Conduct is Contagious."[137] So it proved. Shortly afterward, thirty-seven furnacemen quit the works of the Birmingham Copper Company.[138] In such circumstances, copper masters reached for legal handbooks like Burn's *The Justice of the Peace and Parish Officer*, from which they could pluck the powers best suited to their purpose. John Vivian urged his son, J. H. Vivian, who had day-to-day management at Hafod, to acquaint himself thoroughly with the disciplinary resources at his disposal: "It is worth yr while to consider, in what your Men are subject to the Control of Magistrates—If they quit their Work, no doubt, they come under the 20th and 21st Geo. 2d—Or, if they combine, so to do, they subject themselves to Indictment."[139] What would be tactically best, in other words? Should Hafod workers be pursued for breach of contract, or indicted for conspiracy? J. H. Vivian was left to mull it over. Of one thing, however, his father John Vivian was certain: "I think the men who broke the Prison should be indicted as Rioters," a capital offense.

The Threefold Crisis of the Early 1840s

Workplace friction was a structural feature of early industrial capitalism. It could not be otherwise in an age of dizzying change, when once-familiar social coordinates seemed to vanish, and when economic conditions swung wildly from boom to bust. Yet, as we have seen, some commentators argued that the copper industry was unusually free of conflict. The *Morning Chronicle*'s correspondent in 1851 drew a contrast between the copper business and the notoriously unstable iron industry: "Copper being an expensive metal, and not so much in demand for the rough purposes of commerce as iron is, does not fluctuate in value according to the prosperity or depression of the times in a like degree with iron." Moreover, the copper masters, being oligopolistic, were able to keep volatility in check: "The copper trade is in a very few hands, and these acting in concert, can better regulate prices and keep them at such a rate as to allow of good wages; and it has always been the endeavour of the copper masters, though every change of circumstances in their trade, to keep their wages as equal, year after year, as they possibly can, in which they have so far succeeded that the changes have been very few." The outcome was a happy one: "The son succeeds his father in the works, and lives his life out, without apprehension of change or discomfort arising from adverse times."[140]

This was altogether too roseate a view. Disputes were not aberrant; they were, in fact, endemic. This was true at all times; it was especially true at times of rapid growth, when furnaces were being enlarged and workshops extended. Amid the tumult of building and rebuilding, new optima for surplus extraction had to be established. At such times, new divisions of labor had to be devised and heightened workloads imposed. Managers had to ensure that the return on capital investment went to capital, wholly to capital, not to labor. Their greatest fear was a "compact" among the furnacemen, an unspoken agreement to put a ceiling on output. William Jones, the manager at Hafod in the 1830s, was convinced that his workers had decided to make no more than six tons daily. In August 1834, he found the furnacemen "indifferent about aiming at full work." Nevertheless, "as we get a few 7 tons Daily," Jones in this instance attributed sluggishness on the part of furnacemen to the extreme summer heat rather than a "compact to bring down the Works to 6 Tons." The threat of a go-slow was ever-present, though, "which we must endeavour to avert by every possible means."[141] Just weeks later, William Jones noted that "a perfect indifference to full work and indeed a manifest disposition not to do it" was once again on show at Hafod, among the younger men in particular. He detected it in "a second rate workman named Enoch Thomas, who bullied another . . . of doing full work." Enoch Thomas was immediately discharged ("which I believe had some effect on other youngsters").[142]

William Jones's conviction that the furnacemen at Hafod were curbing output was not a piece of paranoia. Limiting production was a perfectly understandable response to the ratcheting-up of workloads. Furnacemen were paid by the ton, but much depended on how the ton was defined. When the employers sold copper, they did so in tons of twenty hundredweight (equal to 2,240 pounds). The unit of measurement within the works, however, was a ton of twenty-one hundredweight (equal to 2,352 pounds). In other words, furnacemen were required to handle a ton that was 5 percent heavier than the standard. Employers rationalized this by claiming that the extra 5 percent represented the weight of moisture in the ore, plus the weight of the box in which the ore was carried. That much was an accepted convention in the trade. Attempts by employers to increase the weight of the workmen's ton to twenty-two, twenty-three, or even twenty-four hundredweight were less acceptable. Yet those attempts were made in the early 1840s when three convergent crises affected the Swansea District.

The first of these was global; it was the first great cyclical crisis of industrial capitalism as a global system. Financial institutions seeking an outlet for overaccumulated capital catalyzed the crisis. Investable funds were put in pursuit of viable opportunities. When the most plausible options were exhausted, attention turned to ventures in which the returns were far from certain, and then, when even they were fully subscribed, to speculations that were frankly hazardous. The money that flowed, for want of an alternative, into rickety, overextended operations produced losses that destabilized credit more widely. Industrial restructuring and mechanization aggravated the crisis. In Britain, the immiseration of handloom weavers, driven to economic extinction by the take-up of power-loom weaving, stands out as the era's most grievous instance of technological unemployment, yet it was not alone. At the turn of the 1840s, one sector after another toppled into a spiral of falling prices and contracting demand. Copper could not escape the downward gravitational pull. In 1837, British copper production stood at 15,350 tons; by 1842, it had slumped to 9,940 tons.[143] The price of copper on the British market followed suit, sinking to levels not seen since the 1780s.[144]

The second crisis was political, part of the complex transition to free trade that the British state negotiated in the 1840s in response to the inaugural crisis of industrial capitalism. The budget introduced by Sir Robert Peel in 1842 was an attempt to alleviate the unprecedented economic downturn of the five preceding years and the acute social crisis it precipitated. The 1842 budget's fundamental principle was the reduction of duties, especially those on industrial raw materials. This principle was not applied to copper ore, however. Strenuous lobbying by Cornish MPs had the duty on overseas ore *increased*, and the privilege of smelting in bond withdrawn on the grounds that it afforded foreign manufacturers access to British-made copper at bargain prices—prices not available to their British counterparts. Cuban and Chilean ores, the rich oxides and carbonates used in operation IV of the Welsh Process, which had been a prominent feature of the Swansea scene since the late 1820s, were now subject to import duties, and heavy ones at that.

Smelters' costs were rising. They were driven further upward by the third crisis, one local to the copper trade in South Wales. Keeping down the cost of furnace stuff was a perennial concern of the smelting companies. As a rule, they were able to do so. They were few in number and operated cheek by jowl within a ten-mile radius of Swansea. The mine proprietors from whom they bought, by contrast, were numerous and scattered. It was a relatively simple matter, therefore, for the smelters to act in concert and impose their wishes on the sellers of ore. Sometimes they were brazen about it, joining together in formal price-fixing organizations, such as the Copper Trade Association of the 1820s. At other times, a looser form of collusion sufficed, organized via the regular ticketings held at Swansea. In 1841, however, an aggressive new entrant, the English Copper Company, refused to abide with the arrangements that put a ceiling on the cost of furnace stuff.[145] A bidding war ensued.

Caught between rising costs and shrinking revenue, Swansea's copper companies sought to resolve the crisis by cutting the one cost over which they exercised direct control, that of labor. This could be done in two ways—by lowering wages or increasing workloads, or by combining the two in a pincer-like attack on the workforce. Stepping up workloads was best achieved by increasing the weight of material in the standard charge issued to furnacemen, as has already been suggested. All furnacemen, other than calciners, were paid by the weight of the charge. If the piece rate remained unchanged, but the charge grew bulkier, employers could slice into the real cost of labor. This scheme had clearly been practiced in the period leading up to the strike of 1843. The industry-standard ton of twenty-one hundredweight, so Hafod workmen complained, was no more than a memory: the "ton" worked up by ore smelters (operation II) now consisted of twenty-five hundredweight, while slagmen (operation VI) had to deal with an outsize "ton" of twenty-eight hundredweight.[146] In July 1843, the copper masters took further measures: they announced across-the-board wage cuts. The reductions were considerable. The smelters who made "white metal" (operation IV), the group composed of "the most moral intelligent workmen" according to Le Play, were to be docked 12.5 percent of their wages. The roasters who handled operations VII, VIII, and IX were to lose 12.7 percent. Ore smelters (operation II) stood to forfeit 13.5 percent. Slagmen were to suffer the most: a 25 percent cut from 30s. per week to 22s. 6d.[147]

Strike action ensued, involving furnacemen at all the works in the Swansea, Neath, and Afan Valleys. The works affected along the Tawe were those of the Vivians at Hafod, the Grenfells at Middle Bank, the Forest works of Benson, Logan & Co., the White Rock works of John Freeman & Copper Co., and the various works operated by Williams, Foster & Co.—Morfa, Landore, and Rose. To the east, there were walkouts at the Neath Valley works of the Crown Copper Co. and the Mines Royal Society. Still further to the east, strike action hit the Cwmavon works of the English Copper Co., and the Vivians' junior works at Margam. Only the west of the Swansea District, where the Llanelly Copper Company was the sole survivor of the tough trading conditions that had shuttered the Cambrian and Spitty Bank works, appears to have been unaffected.

That the employers should have proposed cuts of this magnitude in the summer of 1843 speaks to the gravity of the crisis—the multiple crises—afflicting the industry. The wage reductions were provocative, and yet the employers pressed ahead with them at a time when southwest Wales was already in uproar. In the rural districts to the west of Swansea law and order had been set at naught by an agrarian protest movement. The so-called Rebecca Riots had begun at the end of the 1830s; in 1843, they were at their height. In essence, the "Rebecca-ites" were in revolt against the capitalist modernization of their impoverished communities. The small farmers who provided the bulk of Rebecca's followers were particularly aggrieved by ambitious development projects, turnpike roads especially, that paid dividends to wealthy investors while putting obstacles, literal obstacles in the form of tollgates, in the way of poor country folk taking what little they could spare to market. "Rebecca's Children" refused to accept that the public highway should be a source of private gain and set themselves the task of destroying the tollgates that had spread like a rash across southwest Wales in the 1820s and 1830s. Armed parties, their faces blacked, roamed the countryside at night, smashing gates and demolishing tollhouses. They invoked scriptural authority for their actions, citing the Book of Genesis, which prophesied that the descendants of Rebecca, the wife of Isaac, would "possess the gates of those which hate them."[148]

These agrarian redressers did not stop at leveling tollgates; their grievances were many. Being drawn from a population that dissented from the Established Church, they seethed at the payment of tithes to support the Establishment, especially when those tithes were intercepted ("impropriated") by a lay patron from the landlord class. Indeed, Rebecca's followers struck back against all manner of gentry-sponsored "improvements." They demanded unimpeded access to common lands; they called for a reduction in exorbitant rents. Landlords who ignored these calls could expect retribution. Under the cover of darkness, hayricks would be torched, ripening crops trampled, and prized orchards hacked down.[149] Magistrates seemed powerless in the face of the insurgency. Indeed, many of them were under siege. Having burned down the tollhouse at Llanon, Carmarthenshire, early in August 1843, a party of Rebecca-ites "then proceeded to a mansion recently erected near there by Mr. Rees Goring Thomas, a magistrate of the county, and broke all the windows."[150] Rebecca had good reason to target Rees Goring Thomas. He was no rustic squire; he was a London financier, a proponent of highway improvement (especially in the neighborhood of his own estates), and a much-loathed impropriator of tithes.[151] He was also closely connected to the copper industry. As we have seen, Goring Thomas was a shareholder in the Cobre Mining Association, and two of his sons were partners in the Llanelly Copper Company.[152] He revealed how the eastern portion of Rebecca's realm and the western part of the Swansea District overlapped.

The New Poor Law of 1834 was another grievance taken up by the Rebecca-ites. Their hostility to the 1834 act and its policy of incarcerating the needy within "Poor Law Bastilles" was illustrated to dramatic effect on June 19, 1843, when hundreds of protesters descended on Carmarthen. The townspeople joined with them to sack the recently built workhouse. It is likely that the building would have been razed completely had not a detachment of dragoons clattered into the town and chased off the crowds. As the summer advanced, rumors started to fly that Swansea was to receive a similar visitation: "Great alarm has been excited in Swansea and its neighbourhood by a report industriously circulated there that a procession of Rebeccaites, similar to that which paraded Carmarthen on the 19th of June, is to visit the town of Swansea in the course of the week."[153] Ominously, the descent of the Rebecca-ites would coincide with the introduction of the new pay scales at the copper works at the start of August: "What tends considerably to heighten the alarm is the fact, that in consequence of the reduction in wages by the masters in the various Copper Works which has taken place during the past week (in many instances to a very considerable extent), numbers—nay, hundreds of workmen have refused to accept employment unless at the old rate of prices, and have therefore joined the ranks of the disaffected. Those copper-men are to form a principal feature in the contemplated procession."[154]

In the event, the Rebecca-ite hosts failed to appear, mindful, no doubt, that one hundred men of the 75th Regiment had been hurriedly billeted in the town; but the copper workers did assemble for a show of strength. A body of men, from one thousand to fifteen hundred in number, marched on Swansea. They were met, before reaching the town, by the mayor, a party of police, and several employers. J. H. Vivian, the manager of the Hafod works, persuaded the strikers to retrace their steps to a field above Hafod for an open-air conference. *The Times* reported on the exchange that followed: "'And now,' said Mr Vivian, 'if there is any man who has any grievances, let him come forward and state them.' After a minute or so, a workman stepped forward and said—'We want our price, as we had it before.'"[155]

Vivian would have none of it: "We cannot give more. It is a matter of necessity." Besides, were not the coppermen, even on the reduced pay scale, much better off than their wretched rural cousins? The workmen thought not. They complained that they could seldom, if ever, earn the wages the masters claimed were theirs. Sickness and deductions saw to that. Moreover, the nature of their work demanded a special diet. The "amount of bread, bacon, and cabbage, on which the farm servant lived and grew fat, would not suffice to keep them alive." Working a furnace required a prodigious calorific intake, and the wage reduction imposed by the masters would not allow of it, hence the walkout.[156] "We would rather starve idle than starve while working hard," one worker snapped at Vivian, "and that's an end of it."[157]

The furnacemen's rhetorical strategy was to present their work as uniquely demanding. They shouldered burdens that were quite beyond the ordinary, yet they only earned wages commensurate with their superhuman workloads after many years in the post. For their part, the masters liked to portray themselves as paternal but respectful of their employees' independence. The Swansea District, they pointed out, had no truck system, with men paid in tokens that could only be redeemed at a company store; men were paid weekly in hard cash, allowing them to pay out for the protein-rich diet to which they attached so much importance ("savoury dinners of breast of mutton, turnips and potatoes" delivered to the works by wives and daughters).[158] By striking, the furnacemen were rejecting the parental solicitude that had always been extended to them. The masters made much of their wounded feelings. "I have always identified myself with the workmen," said J. H. Vivian; "[and] I must express my surprise that my own men . . . should join the others in the strike." It was black ingratitude—all the blacker, in fact, once the parlous state of the industry was taken into account. For Vivian and his fellow employers to keep their works open in the circumstances of the early 1840s was, they claimed, nothing less than philanthropy: "We, as masters, care but little about carrying on the works, which have been carried on without profit for this long time."[159]

Yet the strike was more than a squabble about cash. For the masters, the dispute was an opportunity to resolve a crisis in which questions of local authority and international competitiveness were intertwined. For the furnacemen, the strike was an attempt to arrest a remorseless increase in workloads. It was also a protest against the seniority system that had long prevailed in the industry. Although many contemporaries thought advancement by seniority made for industrial tranquility, early Victorian furnacemen demurred. Two workmen, John Evans and David John, testified to that in a letter published in the *Swansea Journal* in the strike's first week.[160] "This," they told readers, "is to show you the state of the coppermen and how they are used at the copper works." Claims about the extravagant wages paid to furnacemen, made repeatedly by the masters and parroted uncritically in the local press, ignored the long years of poorly remunerated work that awaited young men joining the industry: "When they take a workmen in to the copper works he must work for from 1s. 6d. to 1s. 8d. a day, and he must be like a hackney about the works perhaps for a year before he can have a calciner's place; and he must work as a calciner for the lowest wages for many years before he can have the chance to have a whole furnace." Advancement was slow. For some furnacemen, it never came. The point was underlined by some Hafod workmen in a letter to *The Cambrian*: "We would appeal to those who look with a magnifying eye upon the high rate of wages, to consider impartially but for moment the different hard processes which coppermen are gradually going through before they reach these high wages; some 10, 20, 30, and even 40 years in the works before they obtain but a very insignificant advance, so that but a few ever reach the highest wages, while the majority have worn out their strength and lives in the bloom of their days."[161]

Forcing the masters to withdraw the new pay scales would not be easy, however. It would require unity on the part of furnacemen across the Swansea District, a unity that was hard to realize. The employers had been careful *not* to impose a blanket reduction on their workers; the wage cuts varied widely. As we have seen, the percentage lost by slagmen was double that forfeited by metal smelters. The copper masters liked to present their approach as driven by a desire for equity. "You know," J. H. Vivian declared, "that it has been considered that the work at the metal-furnaces is not so hard as at the ore-furnaces; and, therefore, those men had been reduced a little more than the men at the ore-furnaces."[162] It is difficult to believe, though, that the asymmetrical reductions were not tactical, aimed at dividing the workforce. If so, the tactics paid off. The strike was never completely solid. Most of the works in the Lower Swansea Valley turned out on August 3. Men from the Forest works joined the strike a day later. Furnacemen in the Neath Valley were also quick to down tools. Their counterparts at Cwmavon, on the other hand, waited until August 19.[163]

Maintaining discipline was a daunting challenge in a dispute that involved workers from so many different plants. The *Morning Chronicle* detected divisions in the furnacemen's ranks within a week of the strike beginning: "The coppermen still express their determination to 'hold out to the last'; but we suspect, from the system of surveillance pursued towards some of their number, especially among the Forest men, that there already exists a great difference of opinion between them."[164] Indeed, a few men at the Middle Bank works appear to have worked throughout, an example that proved fatal to the strikers' cause. It led a handful of men to recommence work at the Hafod works in the last week of August, prompting a night of violence in neighboring Landore, where the scabs lived. The next day, hundreds of strikers gathered outside the Middle Bank works "in a menacing attitude," determined to picket out those at work lest the contagion spread further. Magistrates invoked the Riot Act to force their withdrawal.[165] By early September, matters were becoming desperate for those on strike. *The Cambrian* reported that the "expedients to which the men have had recourse, in order to support their families during the stoppage of the Works, especially the dividing of the funds of the clubs, and the pledging of articles of clothing, &c," could go no further.[166]

Defeat

On Thursday, September 7, the Hafod men agreed to return to work "at the reduced rate of wages originally proposed by their employers, Messrs. Vivian, after a *strike* which has lasted for five weeks."[167] The resumption of work at Hafod, one of the Swansea District's largest works, triggered capitulation elsewhere, as *The Cambrian* reported:

> Since writing the foregoing, we have learned, that the men belonging to the Forest *Copper* Works returned to their work last evening, at the reduced wages. We find also, that most of the Middle Bank coppermen have applied to be received back, and that Messrs. Grenfells have now as many men as they require for carrying on their works. The Upper Bank *Copper* Works are still idle, and we understand that Messrs. Grenfell are not anxious to light these works. We hope in our next to be able to announce that the men in the employment of Messrs. Williams, Foster, and Co., Freeman and Co., and the Mines Royal Co., have likewise resumed working, and that the whole district may again be in operation.[168]

The Cambrian's hopes were realized; work resumed across the Swansea District on Monday, September 11, 1843.[169]

It was an emphatic defeat for the furnacemen, one that allowed the masters to escape the entangled crises of the early 1840s on the terms they wanted: wages were slashed and excess capacity sloughed off. Having lowered labor costs and trimmed workforces, the masters turned their attention to regulating the price of ore. The market for Cornish ores had been unsettled since the English Copper Company began its disruptive bidding war at the start of the 1840s. The formation of a new Copper Trade Association in 1844 was a sign of renewed collegiality.[170] Collusion could reign once more. The wider social context also took on a more tranquil aspect. After 1843, a mixture of military force and political concession took the sting out of Rebecca, while Chartism, the movement for democratic reform, which had been a conspicuously "physical force" in South Wales, took a less insurrectionary road in the mid-1840s. The masters had every reason to be happy with the outcome of 1843. The first district-wide strike in the industry was also the last.

For furnacemen who had exhausted their savings and whose furniture and Sunday-best suits were in the pawnshop, the masters' terms were hard to accept. Indeed, resentment flared up almost as soon as workers had settled back to work, as *The Cambrian* reported: "The town of Neath was thrown into some excitement on Monday afternoon, in consequence of its being rumoured that several of the workmen belonging to the Crown *Copper* Works had been apprehended for refusing to proceed with their work, which they had but recently resumed, after a *strike* for wages which had continued for several weeks."[171] Local magistrates brought an end to this dispute by threatening the strikers with jail. If, the magistrates continued, the men were "dissatisfied either with the nature of the work they had to perform, or the amount of remuneration," they should see their contracts out "in a legal manner" and then take themselves off. The opportunity of taking themselves off was, in fact, one that a growing number of workmen embraced in the years following the 1843 strike. The masters' triumph was not as complete as first it seemed. The Swansea District's brief moment as a global hegemon was drawing to a close. Henceforth, those versed in the Welsh Process would be able to find employment in copper smelting settlements in the Americas, Australasia, and southern Africa. Out-migration began. The numbers were not large at first, but they heralded a new division of labor within the global copper trade.

CHAPTER FIVE

Global Fragmentation, 1843–c. 1870

The period between the late 1820s and the early 1840s saw Swansea's influence over the global copper trade at its zenith. Thanks to their ability to smelt in bond, works in the Swansea District were able to draw upon ores from the Caribbean and Latin America. Through the 1830s, therefore, South Wales exerted a powerful centripetal pull on distant mining districts; it became the "central smelter for minerals from East and West." From the mid-1840s onward, though, that centrality became less pronounced. The Swansea District continued to exercise a global influence, but it was no longer the "central smelter." It provided instead a model for new centers of smelting to emulate. The centripetal energy generated in South Wales still drew in furnace stuff from far and wide, but there was also a countervailing centrifugal push sending personnel and techniques outward. Furnacemen and specialized industrial equipment left the Swansea District for destinations in Chile, Australia, and North America. The Swansea-centric world of the 1830s was increasingly decentralized by the 1850s.

This shift had its origins in the disordered conditions of the early 1840s. Economic depression dampened demand; the masters' attempts to drive down costs provoked workplace militancy; and the collusion between masters that had long kept ore prices in check collapsed. By the mid-1840s it seemed as though that crisis had passed. There were signs of economic recovery, albeit fitful. The strike of 1843 had ended in bitter defeat for the furnacemen, enabling copper companies' margins to be restored at the cost of worker impoverishment. Finally, order was restored (as the Swansea copper companies saw it) in the market for furnace stuff; a new Copper Trade Association was able to put a ceiling on ore prices. Paradoxically, the effects were not entirely to Swansea's advantage. Resolving the crisis of the early 1840s proved as damaging as the crisis itself. This became apparent as ore prices internationally began to drift downward.

Central Smelter No More: Chile

Swansea's new difficulties first manifested themselves in Chile. The high ore prices of the early 1840s had been a boon to Chilean exporters; the slide in prices that followed was therefore keenly felt. Every shilling that came off the price of copper ore at the Swansea ticketings was a shilling lost to shipping agents in Copiapó or Huasco and to mine owners across the Norte Chico. And as prices fell lower the virtues of retaining the ore for smelting in Chile became more and more apparent. Changes to Chile's commercial code between 1845 and 1848 added to the incentive. Duties on imported coal were reduced, and the restrictions that obliged importing vessels to dock at Valparaíso or Coquimbo before proceeding to ports closer to the mining districts were relaxed.[1] Taken together, these developments led to a surge in Chilean smelting. "In Chili," one British lobby group complained, "the art, which before 1842 had shown symptoms of neglect, has been resumed with new energy, under all the encouragement which the [Chilean] Government can give, and the export of ores thence is gradually yielding to the export of the perfect metal."[2] The challenge to the Welsh copper industry was plain enough.

The threat from Chile came not so much from traditional mining/smelting centers as from an entirely new smelting sector established in imitation of Swansea. The reverberatory furnace had been trialed in Chile in the 1820s by Charles Lambert, but its use was very limited until the mid-1840s. Adapting the reverberatory proved a complex matter, not least because its prodigious appetite for fuel was a significant demerit until coal supplies became more elastic after 1845. Besides, Lambert and his furnaceman (David Lewis, whose name suggests a Welsh origin) concentrated at first on a niche operation. The blast furnaces conventionally employed in Chile were effective at reducing oxide ores; they did not excel at isolating the copper in more complex sulfides. The reverberatory furnace did. It could therefore be applied to the smelting of ores that locals disregarded or to the reworking of copper-rich slag that had been discarded by earlier generations. But until the barriers to the use of imported coal were lowered, the Welsh Process was not widely practiced.

The changed trading conditions of the mid-1840s stimulated the foundation of "English [*sic*] works" worthy of the name: establishments that were significantly larger than traditional Chilean smelting works (although far smaller than the Swansea norm) and that burned coal (in part, at least). Those built by Joaquín Edwards at Lirquén, near the coalfields of Concepción, were authentically "Welsh" in the sense that they were sited near the source of fuel. Edwards and his associates were responding to the law of September 24, 1845, which freed from export duty all copper bars smelted with "native fuel" in the Chilean south.[3] Other works were built on the northern coast, the better to exploit the resources of the Norte Chico, which were then being intensively developed. Joaquín Edwards hedged his bets by taking this option too, establishing a works at the port of Coquimbo. Charles Lambert, meanwhile, built a smelter at Compañía near La Serena, on the wide bay that swept north from Coquimbo. This boasted five reverberatories in 1845, nine in 1847. Working from customs records, Valenzuela estimates that the Compañía works exported about five thousand tons of copper bars in 1845–1850.[4] As early as 1846 the Compañía plant was functioning so well that it was able to turn out bars of refined copper that were equal "in quality shape and appearance" to Best Selected, the premium brand of the Swansea District—or so one spooked Welsh copper master reported.[5] Further investment in the north came from the British-owned Mexican & South American Company, which dispatched a twenty-one-strong advance guard to Chile in 1847: a manager, a chemist, a blacksmith, a carpenter, two masons, a refiner, and fourteen furnacemen. They began the construction of a works at Herradura; additional capacity came afterward at Tongoy and Caldera.

The Mexican & South American Company was a significant new entrant. Other smelting companies that began operating with reverberatories in Chile did not tamper much with the sequence of roastings and tappings that made up the Welsh Process; they experimented with the fuel, sometimes trialing wood, at other times mixing local coals with Welsh imports. The Mexican & South American, on the other hand, looked for a radical shortening of the Welsh Process. It was licensed to use James Napier's patent method of smelting, which promised a sharp abridgment of the traditional Welsh Process. This, depending on whose account was to be believed, might involve six, seven, eight, or ten separate processes. Napier's patent process offered a dramatic abbreviation. With the right inputs, just two stages would suffice. The implications were clear: if fewer separate operations were required to reduce the ore, less fuel was needed. And if that was so, why carry ore to Swansea Bay? And why not bring in Welsh furnacemen, a good many of whom were ready to emigrate in the wake of their defeat in 1843, to work locally? The abolition in 1843 of long-standing bans on the emigration of skilled workers and export of specialized equipment from Britain opened the door to just that possibility.[6]

Central Smelter No More: Australia

The prospect of bringing in Welsh furnacemen was also being broached in another part of the southern hemisphere: Australia. A sensational mineral discovery had been made in 1844, offering an economic lifeline to the struggling colony of South Australia. Established just eight years earlier, South Australia was intended as a haven for the overspill population of Britain. These would be free settlers, men and women who sought a new beginning in the Antipodes. They were to be offered a place warranted free of convicts, and therefore free of the risk of moral contamination that association with the shackled and shady denizens of New South Wales surely would have brought. The first settlers were landed and a colonial capital, Adelaide, was marked out, but a sustainable economic role for the colony was harder to settle on. By 1840 the whole venture was tottering on the verge of bankruptcy. Salvation came in the form of metals. Soon after the foundation of the colony, prospectors, often experienced Cornish mine captains, moved into the Mount Lofty Ranges that rose east of Adelaide. Before long the first mineral strikes were made and telltale Cornish toponyms began to dot the map. In 1841 Wheal Gawler began operations as Australia's first metalliferous mine. It yielded silver-lead ore, but many of the mines that followed soon afterward produced copper.

The first major discovery beyond the Mount Lofty Ranges was at Kapunda, fifty miles northeast of Adelaide. The promoters were able to secure eighty acres of unsurveyed Crown land at a bargain rate. It made their fortune, for the carbonate ores that outcropped on the surface were assayed at 22.5 percent copper, and the zone of enriched ore, it was subsequently established, extended to a depth of three hundred feet. Mining began in 1844, employing familiar Cornish methods and described in the distinctive terminology of Cornwall's miners.[7] "Tutworkers" hacked through the non-mineralized rock, exposing mineral veins for "tributers" to work at. The ore brought to the surface was said to be "at grass." A painting of the surface workings from 1845 shows a horse whim in use to raise water (figure 5.1). The long wooden ore barrows beloved of Cornish miners lie around. A beam engine, that unmistakable visual signifier of Cornish mining practice, was to be erected shortly afterward. The ore was carted to Port Adelaide for shipment to Swansea.

Figure 5.1. Kapunda Mine, by S. T. Gill, 1845
Watercolor on paper, 27.6 × 40.0 cm.
Gift of the South Australian Company 1931
Art Gallery of South Australia, Adelaide
0.942

No sooner had Kapunda gone into production than a still more notable mineral strike was made farther north. In 1845 pastoralists reported outcrops of very rich ores (azurite, malachite, and cuprite) alongside Burra Creek. Assaying revealed the superior ores to have a truly head-turning metallic content. The very best were assayed at 44 percent.[8] A bonanza was in prospect and local investors scrambled to find the necessary development funds. A special survey would be required, for which the colonial government would charge an exorbitant fee. Two partnerships came together to split the cost and to divide the mineral field into two lots. The "Snobs," an alliance of Adelaide shopkeepers who gathered in the South Australian Mining Association (SAMA), took the more northerly; the "Nobs," the landowners who formed the Princess Royal Mining Company, took the area to the south. It was the Snobs of the SAMA who struck lucky. The ore deposits allotted to the Nobs proved superficial; those scooped by the SAMA were both fabulously rich and extensive.

Having taken possession of its lot, the South Australian Mining Association took the obvious next step. It hired a Cornish mine captain, Thomas Roberts; he, in turn, recruited a party of fellow Cornishmen. (The feast of Piran, patron-saint of Cornish miners, was celebrated at Burra in 1848, the *South Australian* reported, with "wrestling matches, the favourite amusement of Cornwall.")[9] When work began in September 1845 Burra was still a frontier spot, but over the next five years more than fifty-four thousand tons of dressed ore would be produced from what was soon known as the "Monster Mine" (figure 5.2). The profits were stupendous, "a sort of El Dorado dream."[10] The SAMA paid fifteen dividends, each of 200 percent, in the first five years of its existence. Not surprisingly, shares in the mine exchanged for dizzying sums. By 1848 shares with a nominal value of £5 were being quoted at £200.[11]

Figure 5.2. Burra Burra Mine, by S. T. Gill, 1850
Watercolor on paper, 32.8 × 57.5 cm, showing chief portion of surface operations. Gill presents a teeming mining landscape, with a Cornish engine house at its center. A year later, with the gold rush in neighboring Victoria drawing miners away, the scene would have been rather less busy.
Gift of Mrs. F. M. Graham and family 1947
Art Gallery of South Australia, Adelaide
0.1352

These riches were made on the back of exporting the ore to Swansea. There was still more to be made, the SAMA directors reasoned, if the ore could be smelted in South Australia. Why should Swansea's copper smelters share in the plenty? Indeed, Australia's subordination to British industrial interests roused the passions of many a budding colonial patriot. The *South Australian Gazette* lamented "the enormous sums lost to the colony on the valuable ores shipped as they have been at high freights and heavy charges to be smelted 16,000 miles away from the place of their production, and half as much more from their ultimate market."[12] The beckoning markets which the *Gazette* had in mind were those of India and China; they were markets that South Australia was well placed to serve, if only smelters could be built in the colony. Happily, that very consummation was at hand. Several smelters were erected in the late 1840s to serve the Mount Lofty mines and to receive ore brought down from Kapunda. A Welsh-style reverberatory was trialed at Bremer in the Adelaide Hills in 1848, but it was not, at this formative stage, certain that smelting in the Welsh fashion would prevail. The proprietors of the works at Yatala near Port Adelaide, for example, opted for a blast furnace.[13]

The SAMA's initial venture into smelting was not down the Welsh route. The directors hired a clutch of German technicians to build a continental-style blast furnace adjacent to the Burra mine. Nothing came of it. The furnace was small and underpowered, and the experiment was abandoned in 1847 without any tangible outcome. The SAMA directors remained set on smelting their ore locally, but now they turned to the Welsh Process; or rather, they turned to James Napier's much-shortened version of it. The rights to Napier's patent were held by the Patent Copper Company (PCC), which had recently bought the Spitty Bank works on the Swansea District's western edge. The PCC was an initiative of John Schneider & Co., a London firm heavily involved in mining and metals.[14] The Schneiders were already active in the Mexican & South American Company, which was then trialing Napier's process at Herradura in Chile. The attraction of Napier's process lay in fuel economy, a feature much sought after in the fuel-short Norte Chico. It was a feature that also attracted the attention of the SAMA, which could not access the abundant coal reserves upon which the Welsh Process in its traditional form depended. The SAMA therefore invited the Patent Copper Company to establish a smelting works at Burra. The SAMA would continue to export its premium ores to Swansea, but the PCC works would be supplied with the poorer grades of Burra ore, ones that would not repay the cost of shipment overseas. Agreement was soon reached: the PCC would build a Welsh-style smelting plant at Kooringa, close by the Burra mine. Everything needed, from the firebricks to the workforce, would be shipped in from Wales.

The departure of the first emigrants on the *Richardson*, a 361-ton bark, was reported in *The Cambrian* of Swansea in June 1848: "This splendid vessel, the property of Messrs Leach Richardson & Co left this port on Monday evening bound for Port Adelaide, Australia, taking with her several cabin passengers, and about 70 steerage passengers. They were chiefly parties from this neighbourhood who are engaged to be employed in a smelting establishment in Australia, under the direction of Mr Tho Williams." Indeed, steerage was occupied almost exclusively by the twenty-five copper smelters and their families who had pledged themselves to the PCC for a seven-year term. The *Richardson* was the Mayflower of Australian copper.[15] Many of those who sailed on her were to play a decisive role in Australian industrial development in the decades to come.

The Kooringa workforce was drawn from the eastern and western extremities of the Swansea District—Llanelli in the west and Cwmavon in the east. Recruitment in the Llanelli district was an entirely natural step for the Patent Copper Company; it most likely drew upon its existing workforce at Spitty. New homes were to be built for the emigrants in a township named for the homeland they had left—"Llwchwr" (Welsh for Loughor). The PCC had no obvious link to Cwmavon, but it took little perspicuity to see that the town was thronged with copper workers who might be attracted by a new life in the colonies. The 1840s had been a roller-coaster decade in Cwmavon. A copper works had been built there in 1839; it joined an already operational ironworks and a long-established coal industry. The entire complex came under the control of the English Copper Company. Rapid expansion followed, fueled by heavy borrowing. In 1847, however, a credit squeeze brought the insecure edifice crashing down and the copper works was taken over by the English Copper Company's principal creditor, the Bank of England.[16] In such troubled times there were copper workers enough, many of them smarting after the defeat of the 1843 strike, who were ready to leave Cwmavon behind and head for South Australia.

Three months on board the *Richardson* gave way to a week-long trek into the interior. The twenty-five pioneers and their families were accompanied by bullock drays that carried all the materials brought out in the *Richardson*'s hold: "48,100 fire-bricks, 40 tons fire clay, 300 furnace slabs, 200 furnace doors, 120 furnace bearers, 10 tons sand," and much else besides.[17] They arrived at Burra at the end of November 1848; by March 1849 smelting operations were underway. A long, open-sided smelting hall accommodated six reverberatories, all of which were soon at work. Indeed, they were soon added to. The workforce was extended too. A further twenty-seven Welsh copper workers arrived on board the *Glen Huntley* in December 1849. The Welsh era in Australian copper had begun.

When a reporter for the *South Australian Register* visited Burra in February 1851 he found two furnace halls and a refinery in full operation: "furnaces roaring, ores roasting, copper and slag cooling, and men stoking and poking on every side."[18] It was a Welsh scene in all particulars save one. The fuel being loaded into the "mouths of the devouring monsters" was not coal; it was wood. Indeed, the reporter was struck by the volume of timber involved: "drays, loaded with wood, were arriving every instant, and were seen from every distance, to replenish the forest of cut logs and long wood piled up, circling the whole area of the works, and filling every available space." Some imported coal was carted up from Port Wakefield at the head of the St. Vincent Gulf, but most of the energy needs of the PCC's Kooringa works were satisfied by burning vegetable matter. From the vantage point of Swansea this seemed a retrograde step, a reversal of the historic switch to mineral fuel that British smelters had made in the seventeenth century. In Australian conditions, however, the use of firewood made sense. The location was landlocked. While the best grades of ore were sufficiently rich to bear the cost of carriage to the coast the lesser grades were not. They had to be smelted at the mine or discarded. Since the lower-range ores contained from 15 to 25 percent copper (more than would be found in most Cornish ores of the time) they were too valuable to be thrown aside. It was these ores that the PCC was contracted to smelt. Perhaps most importantly of all, the country was still very well wooded. In such circumstances the absence of local coal was not a fatal disadvantage.[19]

Burra, like Herradura and the new generation of Chilean smelters, demonstrated that the Welsh Process could be detached from the coal-rich context in which it had first flourished, especially if Napier's patent method was employed. "The system of smelting is the same as that pursued at Swansea," it was said of the works at Burwood, New South Wales, in 1854, "except, that from the favourable peculiarities of Australian ore, three processes, instead of eight, or ten, produce copper, and that of the purest quality."[20] It augured ill for Swansea. If ores could be processed (or partially processed) near the point of their extraction, South Wales's eminence as the "central smelter for minerals from East and West" would come to a close. And that, in essence, was what happened in the third quarter of the nineteenth century. In Australia, the export of furnace stuff to Swansea continued, but it accounted for an ever-smaller proportion of the ore raised at Australian mines. The finest Burra ores were still shipped to Spitty and other Welsh works, but the SAMA was at liberty to have the remainder smelted within the colony. It soon made use of that liberty, sending ores to the nearby Apoinga works of Messrs. Penny & Owens. The discovery of additional ore deposits in South Australia in the 1850s, most notably on the Yorke Peninsula, accentuated the trend. Here, once again, Cornish miners and Welsh smelters were combined to potent effect. The township of Moonta grew into South Australia's "Little Cornwall."[21] The construction of the smelting works at neighboring Wallaroo in 1861 was superintended by Leyshon Jones, once of Cwmavon, who had arrived in the colony on board the *Richardson* in 1848.[22] The works was hailed as "the most extensive in the colony, and the largest, I believe, out of Swansea."[23] Wallaroo had four calciners, twenty-two reverberatories in its main smelting hall, which was 695 feet in length, a separate refinery, and a chimney stack ("a fine pile of bricks and mortar") that was 120 feet in height.[24] The international copper industry, which had been centralized in Swansea for a generation, was fragmenting fast.

Swansea Copper in the 1850s: A Precarious Hegemony

Fragmentation was not immediately apparent in Swansea. The 1850s, after all, saw South Wales acquire a still wider range of mineral tributaries, and, for the first time, dock facilities to accommodate incoming traffic. The copper masters had been bent on harbor improvement since the start of the 1830s, when ore barks from the Caribbean and Chile began to arrive in greater number.[25] "From the extended trade in foreign copper ore." a correspondent to *The Cambrian* argued, "it is of the first consequence to get a secure place for the ships of large burden to rest in."[26] In the absence of a floating harbor, the masters of these vessels had little choice but to ground them near the mouth of the Tawe at low tide, often sustaining damage to their hulls when doing so. Swansea's leading copper masters agitated for an extra 1d. per ton in harbor charges to help fund new dock accommodation.[27] But their enthusiasm was not shared by others. Representatives of the coal trade objected to paying for improvements that, by providing berths for large ocean-going barks, would primarily benefit the copper trade. Coal shippers were happy for additional charges to be imposed on copper and imported ores, but not on coals. Extra expenses, they said, could not be borne in the price-sensitive markets they served, where "a very small increase of charges may wholly shift the course of trade."[28] Wrangling of this kind stood in the way of dock development. Frustratingly for Swansea's copper companies, there were no such delays at other Welsh ports on the Bristol Channel coast. Construction work on a new dock at Newport began in 1835, and in 1839 the new Bute Dock at Cardiff opened for business. "All the other ports in the channel are being improved," J. H. Vivian complained in a letter to Swansea's Harbour Trustees; ". . . we must, to hold our position, improve also."[29]

By the mid-1840s, the matter was urgent. The construction of the South Wales Railway (SWR), laying down a direct link between South Wales and London, had a threatening aspect. The SWR would provide more than a mainline to London; it would connect coastal towns in South Wales. Once the railway was complete, Swansea's mayor warned the Harbour Trustees in 1848, "many large vessels laden with copper ore from Australia and other places, would prefer going to Cardiff, rather than take the ground here, and would send their cargoes to Swansea by railway, so that Cardiff would be the shipping port of Swansea."[30] Happily, a solution was at hand. The Harbour Trustees had authorized a scheme to straighten the lower reaches of the Tawe. Driving through the new channel allowed the river's old meander to be converted into a floating harbor: the Town Dock.[31] Work began in 1849 and the Town Dock was completed in 1852. It was timely. The SWR arrived in Swansea in June 1850, vaulting the river by means of a huge wooden viaduct at Landore with a span of 1,897 feet.[32]

Yet the Town Dock could only ever be a stopgap. It did not meet J. H. Vivian's demand, made in 1848, for "a dock of 10 acres" capable of "accommodating 300–400 vessels."[33] No sooner was the Town Dock open than work began on the South Dock, which opened in 1859.[34] It was soon fully utilized. The 1850s, after all, saw Australian ores joined by other debutants. New parts of Africa, the Americas, Australasia, and Europe were drawn into Swansea's orbit. There were some inevitable false starts and blind alleys. Prospecting in the Caribbean found nothing to substitute for the now-flagging Cuban mining sector. The geological surveying of Jamaica yielded next to nothing, while mining in the Virgin Islands promised far more than it delivered.[35] The promoters of the New Virgin Gorda Mining Company in 1845 claimed that preliminary work on their most promising lode had raised ore with a metallic yield ranging from 18 to 24 percent.[36] The return on investment did not match expectations, however. On the other hand, the first of what was to be a steady series of consignments from Namaqualand on the Cape Colony's Atlantic coast arrived in the mid-1850s, the first supplies from Newfoundland followed, and some ore even made its way from landlocked locations such as Ducktown, Tennessee.[37] Not all new material could boast the high copper content of the ores from the Burra or El Cobre. Pyritic ores from around Huelva, on Spain's Atlantic coast, for example, were low grade, but as they yielded both copper and "purple ore" (an iron oxide that could be sold on to iron manufacturers) they were economically viable—sufficiently viable, in fact, to account for 8 percent of British copper ore imports in the 1850s.

Swansea Copper was now at its fullest extent. Far-flung though its transactions had become, they had acquired a certain regularity and routine. Its way stations, from Coquimbo, gateway to the Chilean north, to Port Wakefield, were familiar to a generation of seafarers and itinerant industrial workers. This was a World of Copper whose citizens had acquired a reflexive awareness of their own globality. An inlet on Kawau Island, New Zealand, where copper mining began in 1845, was christened Swansea Bay, and when investors in Tennessee sought to revamp their Ducktown mines in the late 1850s they could think of no better name for themselves than the Burra Burra Copper Company. Indeed, allusion to the now fabled mine in South Australia seems to have been de rigueur among hopeful mine promoters in the 1850s, for the Cape Colony had its own "New Burra Burra Mining Company."[38]

Swansea Copper had shown itself to be supremely adaptable. It proved capable of settling into some varied ecological niches, provided, of course, that the ore was tolerably close to an anchorage that could accommodate Swansea-bound barks. And if ores were rich enough, mining companies were prepared to persevere in some extremely inhospitable locations. Mining in Namaqualand took place one thousand meters above sea level and far inland from a coast that lacked natural harbors and that was plagued by dense fogs rising from the frigid waters of the Benguela current. It was not the only destination to present navigational or climatic challenges. The shuttle back and forth between Swansea Bay and Cuba had its own dangers. Ships' masters had to time departures carefully if they were to skirt hurricane season in the Caribbean. Lingering in Santiago de Cuba was also to be avoided. Yellow fever was rampant and often claimed the lives of visiting seafarers. In Swansea, *The Cambrian* reckoned that "scarcely a vessel arrives in our port from Cuba but has one, two, three, or even four hands dead on board." Santiago's notoriety was such that ship owners paid danger money to Cuba-bound crews. An attempt to withdraw the bonus payments in 1856 was, *The Cambrian* reported, "most resolutely opposed by a large number of the seamen of our port, who paraded the principal streets of the town, accompanied by three or four flags and headed by a Scotchman playing the bagpipes."[39] More perilous still was the run around Cape Horn, the prospect that faced any captain heading for Valparaíso or the ports of the Chilean north. The outward journey involved battling against westerly gales that seldom abated. On the voyage home barks could race ahead of the wind, but they were now laden with an exceptionally dense cargo, one that had to be checked regularly, for if it shifted the vessel would become dangerously unstable.

Wherever Swansea's influence was felt Cornish mining traditions tended to follow. That was certainly the case in Namaqualand. At O'okiep, tutworkers who drove through non-mineralized rock were paid for the volume of material they handled, while the tributers, who extracted the precious ore, were paid, in accordance with Cornish practice, on the value of what they raised.[40] O'okiep was run by the Cape Copper Mining Company, which had strong connections to England's southwest.[41] It was a given that well-rewarded British workers, authentic labor aristocrats, would set the technological parameters. Most of those who worked in Namaqualand's mines, however, lacked the inestimable benefits of white skin and British citizenship. A good deal of the underground work and all surface tasks were performed by "Native labourers": juveniles and women who undertook the laborious task of moving and dressing the ore. The African labor force was drawn from across the south of the continent: "Damaras, Zulus, Hottentots, Bushmen, &c."[42] Their rewards were meager. The African workers, in the lofty opinion of one European observer, were mostly "from tribes unacquainted with civilization" and therefore, when viewed through the racial optic of the time, naturally unindustrious.[43] Stern discipline was necessary if such disadvantages were to be overcome. Happily for the mining companies, draconian legal mechanisms were at their disposal. Slavery had been abolished at the Cape in the 1830s as part and parcel of a program of emancipation in Britain's wider Atlantic empire, yet no sooner had outright slavery been abolished and the enserfment of native peoples moderated than a new legal code was introduced to pinion African laborers to their tasks. Masters and Servants legislation of an unusually restrictive stripe was introduced in 1841, inaugurating a "labour-repressive economy where workers may not have been slaves, but they were certainly not free."[44] Further legislation, enacted in the mid-1850s as the copper mania in Namaqualand gathered strength, handed still more powers to employers.[45] Swansea Copper put in place what would later become a defining feature of the Witwatersrand and of South African mining as a whole—the division between a privileged white industrial elite and a subordinate black proletariat.

Central Smelter No More: The United States

The global copper industry, which could be mapped in the 1830s in strong, bold lines, most of them converging on South Wales, had to be sketched in softer shades by the end of the 1850s. Swansea remained a prominent feature, still perhaps the preeminent feature, but it was becoming one point in an increasingly multipolar landscape. Despite close and enduring links to Swansea, Chile had maintained an autonomous smelting industry and Australia was rapidly developing one. By the 1850s it was also clear that a further challenge to Welsh hegemony was emerging in the United States. Here, as in Chile and Australia, attempts to smelt on the Welsh model began in the mid-1840s. American industrialists needed no introduction to the *processing* of copper. That was well established, for Paul Revere had built America's first rolling mill for the manufacture of ship's sheathing just outside Boston in the very first years of the nineteenth century. *Smelting*, on the other hand, had never developed very far. US manufacturers relied on British imports, even though these were subject to heavy import duties. Given the rapidly growing domestic market within the United States, there was every incentive to develop a homegrown smelting industry. In the mid-1840s serious efforts to do so began in east coast port cities. British smelters and merchants involved in the import of Latin American ores took immediate alarm. They rang the tocsin in an 1847 memorandum to the British Treasury: "In the United States, where heretofore the smelting of copper did not exist . . . two smelting works, upon an extensive scale, have been established, one at Boston and the other at Baltimore, and are now in operation." Already, they continued, "considerable quantities of copper ores have been received into that country from Chili," an ominous feature.[46]

The new works at Boston was that of the Revere Copper Company, which had been engaged in rolling imported copper for over forty years. When establishing a smelting operation at Point Shirley on the outer edge of Boston Harbor, the Revere Company opted for "the German method, with calcination in the open air, and reduction in the small upright blast furnace." The promoters of the newly formed Baltimore & Cuba Copper Company, on the other hand, settled on the "Swansea (Welsh) method of smelting."[47] Indeed, the Baltimore & Cuba Company set out to create a Swansea-in-miniature on a tongue of land in Baltimore Harbor. Unlike their rivals in the Norte Chico and South Australia, where smelting plants were built close to the mines from which they drew their ore, the promoters of the Baltimore concern intended to ship in furnace stuff from distant locations, in the first instance from Cuba, as the company's name suggested.[48] It would be smelted with local (or relatively local) coal.

Establishing a facsimile of Swansea at the head of Chesapeake Bay required the recruitment of workmen schooled in the Welsh Process. Nineteen Welsh-born "copper smelters" feature in the US census returns for Baltimore in 1850, some of whom can be identified in the British census of 1841.[49] The evidence suggests that the Baltimore & Cuba recruited labor right across the Swansea District. Some came from Llanelli in the west. They were almost certainly former employees of the Llanelly Copper Company, for it was R. J. Nevill, the Llanelly manager, who handled remittances from workmen in Baltimore to family members still in Wales.[50] Others came from the east, like Thomas Llewellyn, formerly of Copper Row, Cwmavon. Emigration was very much in the air in late 1840s Cwmavon, as we have seen. Some of Thomas Llewellyn's neighbors were among those who boarded the *Richardson* for South Australia. Other Baltimore & Cuba recruits came from the Lower Swansea Valley itself. The Leyson family, for example, Americanized into the Elishas by a Baltimore census taker who was stumped by the sibilance of their English, once lived in Llansamlet, the parish that ran down the left bank of the Tawe.

Another Welsh-style smelting concern opened at Canton on the opposite side of Baltimore Harbor in 1850. This new outfit, the Baltimore Copper Smelting Co., was the creation of Isaac Tyson Jr., a Quaker businessman who had already amassed a fortune in the mining of chromite (the chief mineral source of the element chromium) in Maryland.[51] Tyson's mining interests were not confined to chromite, however. He was the proprietor of a copper mine in Vermont, and he had sponsored largely abortive efforts to smelt copper ore there using a blast furnace. The building of a works on Baltimore's waterfront represented a new departure: bringing in ore by sea and smelting it Swansea-style. The new company, which amalgamated with the Baltimore & Cuba in 1854 and centralized production on the Canton site, drew in ore from far and wide. Nothing much came from the Caribbean, even though the allusion to Cuba still appeared in the firm's title (Baltimore & Cuba Copper Smelting and Refining Co.). A good deal came from Chile—cuprite from the El Teniente mine. Increasingly large amounts came from North American sources, however, particularly from Ducktown in eastern Tennessee, where ores were discovered in 1847. This remote Appalachian mine was landlocked, of course, but proximity to a railhead in northern Georgia allowed for the export of the ore, some to Swansea but most to Baltimore.[52]

On the eve of the Civil War Baltimore was America's foremost copper-smelting center. It was made so by exiles from the Swansea District. Over sixty copper smelters are identified in an 1860 city directory, some resident on Locust Point, where the first Welsh "colony" had been established in the mid-1840s, but most of them in Canton, lodged in the workers' rowhouses that stretched along South Clinton Street.[53] Almost everyone listed bore a name that spoke of South Wales: Bevan, Evans, Leyshon (spelled variously), Roberts, and Thomas. They formed a nucleus from which an immense industry would grow. Just as the voyagers on the *Richardson* would seed Australia's early copper industry, so the first arrivals in Baltimore would implant expertise that would subsequently be carried onward to the new industrial zones of the American interior. The son of one of the first migrants reminisced in the early twentieth century: "Henry Johns, my father, came to the United States from Swansea, Wales, his birthplace, in 1846, in his twenty-first year, under contract to work at the Baltimore Copper Works, his brother-in-law, David D. Davis, being superintendent of refining there at that time."[54] This Henry Johns soon moved on to work at—indeed, to build—the Pittsburgh Copper & Brass Rolling Works, a plant designed to handle copper mined on the Upper Michigan Peninsula by the Pittsburgh & Boston Mining Company.[55] It was one of the first signs of a drive into the heart of the continent that would soon upend Swansea's already faltering global leadership.

Gathering Crisis

The uncovering of copper deposits along the shores of Lake Superior in the 1840s added to the centrifugal forces that were pulling at Swansea's position as the "central smelter for minerals from East and West." The effects were not instantly felt. Upper Michigan was remote and unpromising territory. Carriage to emergent industrial centers like Pittsburgh was a lengthy and laborious business. After 1850, however, smelting plants began to be built on the shores of the Great Lakes, first at Cleveland, then Detroit, and in 1860 the Portage Lake Smelting Works became the first smelter in the Michigan mining district itself. The effects were now very apparent. Imports of metallic copper to the United States, principally from Britain and Chile, had been growing through the 1840s; now they plunged. Retained imports of copper, valued at $3.5 million in 1855, fell to about $1.75 million in 1859, then to about $1 million in 1860.[56] The Civil War accentuated these trends. The needs of the war economy gave a terrific stimulus to the Michigan district. The postwar contraction, when it came, was one that low-cost producers on the Great Lakes could ride out. International competitors found it far more testing. A rapid-fire reordering of global production was the result. Several mainstay features of the old Swansea-centric world fell victim to the shakeout. Cornish mining, the foundation stone upon which Swansea Copper had first been raised, slumped, never to recover. In the 1840s Cornwall had been at the forefront of world mining, accounting for 80 percent of British output, which was itself close to a quarter of recorded global output. By 1850 production in the west of the county, the old heartland of Cornish mining, was starting to droop, but compensation came from the southeast of the county, where new mines were being sunk. These developments pushed Cornish copper ore production to nearly 185,000 tons in 1854, an all-time high. The late 1850s, however, saw a downward drift, and the 1860s brought collapse. In 1870 just 56,500 tons was raised.[57]

Ireland, a major supplier of ore to Swansea since the 1820s, went the same way. Output tilted downward in the 1860s, never to recover. Wages dipped too, labor relations deteriorated, and miners began to quit. In 1861 over 2,000 people lived in the mining settlements around Bunmahon on the Waterford coast; by 1881 only 599 remained.[58] The Beara peninsula in the far west of County Cork registered a similar decline. With so little prospect of work, many left for Upper Michigan, where they entered into fierce ethnic competition with Cornish migrants.[59] Chilean mining was also hit hard, its problems compounded by the onset of war between Chile and Spain in 1865 and the complete derangement of commerce and credit that followed. The boom in American copper was ruinous for Chilean producers. The US market had absorbed 44 percent of Chile's copper exports in 1844–1849; in the 1860s, it took just 5 percent.[60] Chilean miners joined their Cornish and Irish brethren in their quest for fresh work. Hundreds headed for the newly opened silver mines of Caracoles in Bolivia or the nitrate fields of southern Peru. Others were reabsorbed into the agricultural sector as the Norte Chico ruralized in the last decades of the century.[61]

Even places that had acquired totemic status in the 1840s and 1850s now started to languish. Underground work at Burra's once-garlanded "Monster Mine" was suspended in 1867. The choicest ores had been worked out and the lesser grades, although plentiful and rich enough in mineralogical terms, could no longer be worked viably because of the "low price which has lately ruled in the copper market."[62] It was the moment for journalists to write elegiac pieces on the transitory character of mineral prosperity. "No mineral discovery in the future," wrote one, "can ever produce so profound an effect as did the finding and development of the Burra Burra." In those days hundreds had toiled underground, "one set succeeding the other in perpetual relay," while above ground "nearly as many hundreds were employed in the multiform processes of raising, dressing, and smelting." Then, Burra and the neighboring townships had hummed with activity. After 1867 the town presented a very different aspect: "The places of business lost the bulk of their custom; the streets the bulk of their traffic; the hundreds of little shanties which dotted the hill sides, and which at the best of times were simple and plain even to ugliness, lost their home look through the desertion of their occupants. An appearance of antiquity and the reality of decay settled down upon the whole place."[63]

When work was renewed in 1870 it was on the basis of open-cut extraction rather than the deep mining practiced by the first generation of Cornish migrants. Yet even this failed to pay in the new era of globally depressed prices. The excavations at Burra were mothballed in 1877 and remained so for a generation. Australian mining continued, of course; indeed, it flourished on the Yorke Peninsula and elsewhere, where ores were raised to supply in-house smelting plants like that at Wallaroo. Yet the umbilical link to Swansea, which Burra had embodied, was lost.

Production at El Cobre also came to a halt at the end of the 1860s. In truth, El Cobre's difficulties were of long standing. Cuban copper ore exports had peaked in 1845. The trend thereafter was downward, as fiscal changes in both Britain and Cuba put a check on production. The British tariff system that had shifted in favor of Cuban ore in the late 1820s took a less favorable turn in 1842 when Sir Robert Peel's government rescinded the privilege of importing ore in bond. Just a year later, a 5 percent export duty on ore was introduced in Cuba. The effect of the two actions was to slice into the once bumper profits of the Cobre and Santiago companies.[64] More worrying for the Cuban mine proprietors were the opportunities that opened up for competitors at this juncture: the Chileans, the Australians, the Americans, and others. Cuban difficulties were compounded by a falling off in the quality of El Cobre's ores as the rich carbonates and oxides that had been brought to the surface in the early days gave way to less valuable sulfides at lower depths.

Cuba's share of the British market declined accordingly. In 1840 Cuba accounted for 62 percent of ore imports into Britain, in 1860 just 17 percent.[65]

The Cobre Company was now the only British concern in operation in Cuba. The Royal Santiago Mining Company folded in 1860. Mining, the Santiago's chairman had to remind querulous shareholders as early as 1849, "was but another name for uncertainty."[66] So it proved. The chairman's address to shareholders a year later was a frank admission of failure. There was "no important discovery of ore" to report. Headings "which at one time promised to yield a fair return, were at the present totally unproductive."[67] Dividend payments became a thing of the past. Instead, the directors made repeated calls for further capital. But further expenditure did nothing to alleviate the persistent losses. The final crisis came in 1858 when the engine shaft collapsed, "in consequence of which the mine became full of water and the works were stopped. The company are now without funds, and the shareholders refuse to respond to further calls."[68]

The Cobre Company was more resilient. To compensate for the decline in ore quality a smelting plant was built at the mine in 1862–1863, and a manager from Neath's Red Jacket copper works was appointed to superintend the work.[69] By making a preliminary reduction of the ore in Cuba the Cobre Company could export a regulus with a metallic content of about 35 percent rather than ore containing only 9 or 10 percent metal and thereby economize on freight. One way or another, dividends continued to be paid, but even the Cobre Company was threatened by the downward lurch in copper prices post-1865. The El Cobre anatomized in the Cuban census of 1862 was not too different from the tumescent mining community of twenty years earlier. It was still dominated by enslaved men of unmixed African ancestry (*morenos*), more than 500 of them.[70] The supervisory layer at the mines remained British, as it always had been. "Most of the captains are from England," an American visitor remarked, "and are practical miners, who have learned their business in the mines of Cornwall and Wales."[71] The census counted 97 Britons in 1862, 91 of them men. Chinese indentured laborers were the only new element: 230 *colonos asiáticos* were registered in 1862. By the late 1860s, however, the size and composition of the workforce was radically different. While the number of Chinese laborers remained constant, much of the company's enslaved workforce had been sold off. Just 75 slaves remained. The British cadre had shrunk too—cut in half, in fact. There were only "40 or 50 British subjects" left by 1868.[72] Clearly, there had been a significant contraction in activity. Indeed, in 1865 and 1866 the company had recorded serious losses for the first time, and the directors took out limited liability protection for the "Consolidated Copper Mines of Cobre."[73] The year 1867 brought better results and a restructuring of the company. It proved a false dawn, for the Cobre Company did not see out the 1860s. Yet the end, when it came, was precipitated not by the slide in prices on the international market but by events in the Sierra Maestra itself, where a major revolt against Spanish rule broke out in 1868.[74] The occupation of El Cobre by insurgents, the destruction of its railway link to the coast, and the conscription of the enslaved workers into the rebel army brought mining to an end. The future of the Cobre Company, which had been uncertain even before the outbreak of war, was now settled. Winding-up proceedings began in London in February 1869.[75] In Swansea, nothing remained but to obituarize Cuban copper. *The Cambrian* recalled days past when "a large fleet of vessels was regularly engaged in the copper ore trade" with Santiago de Cuba. No longer: "Now the importation of a cargo of copper ores from Cuba is regarded almost as a novelty, and almost everyone of the vessels which formerly traded to Cuba has been withdrawn and placed upon other stations."[76]

The landscape was shifting. In the aftermath of the American Civil War, the foundations on which Swansea Copper in its global zenith had rested dissolved in the worldwide crisis over copper extraction. A new era in which Swansea's influence dwindled dramatically now beckoned.

The End of Swansea Copper, c. 1870–1924

Swansea's reign as the dominant player in the world copper trade was a short one. The gradual fragmentation of its mid-nineteenth-century global dominance, visible by the 1860s, became more firmly ingrained over the next three decades. From the 1890s onward its decline was "absolute."[1] The number of firms engaged in copper smelting in the Swansea District fell from fifteen in 1860 to just three by 1920.[2] The technology of the reverberatory furnace, on which rested Swansea's primacy as a smelting center, had been superseded by new metallurgical processes, and contemporaries were aware that it was the end of an era. In 1917, in the first of two lectures delivered to the Royal Society of Arts, Henry Cort Carpenter, professor of metallurgy at the Royal School of Mines, pronounced the Welsh Process to be "now obsolete."[3] An internal report conducted for one of the surviving copper firms in the town in 1924 came to the stark conclusion that "Swansea is badly situated as a centre of the copper trade today" and contemplated the removal of its works to another part of the country or overseas, "where the conditions would enable us to compete for the world's trade."[4] By any measure it was a startling change of fortune for a district that had used its locational advantages to dominate the copper industry for so long, and all the more painful because Swansea's demise took place at a time when world demand for copper was growing. New applications in the electrical industry, in the construction of engines, and other parts for automobiles and locomotives all stimulated growth in output from mines, smelters, and mills. The juxtaposition of a buoyant world market and a declining industry in Swansea has led some commentators to conclude that the Swansea firms must have been guilty of serious business failings to have been unable to capitalize on this situation.[5] For Alfred Chandler, the Vivians and the other British copper firms failed to take advantage of the new technologies available to them and lost crucial market share, with the result that "by 1900 the opportunity for a British firm to become a major player in the global copper oligopoly was gone, never to return."[6]

A full understanding of why the reign of Swansea Copper had run its course by the early twentieth century requires consideration of a complex range of factors, at least some of which lay beyond the control of the district's once-dominant copper firms. The increasing prevalence of new types of copper ore that did not require full reverberatory furnace treatment undid the geographical advantage that the coal-blessed port of Swansea had always enjoyed in the smelting of rich, sulfide ores that needed fuel-intensive refining. The consequences of this were visible not just in the growth of mining and smelting overseas (figure 6.1), but also in a geographical shift in the industry in Britain, where the business of metal refining was drawn into the orbit of the growing chemicals industry, especially in the north of England. In Swansea, firms looked increasingly to the production of other metals and alloys, as well as manufactured goods, to sustain them in business, but this brought them into direct competition with specialist fabricators with better links to the rapidly evolving markets. The needs of the electrical industry, in particular, placed entirely new demands on copper producers to improve the

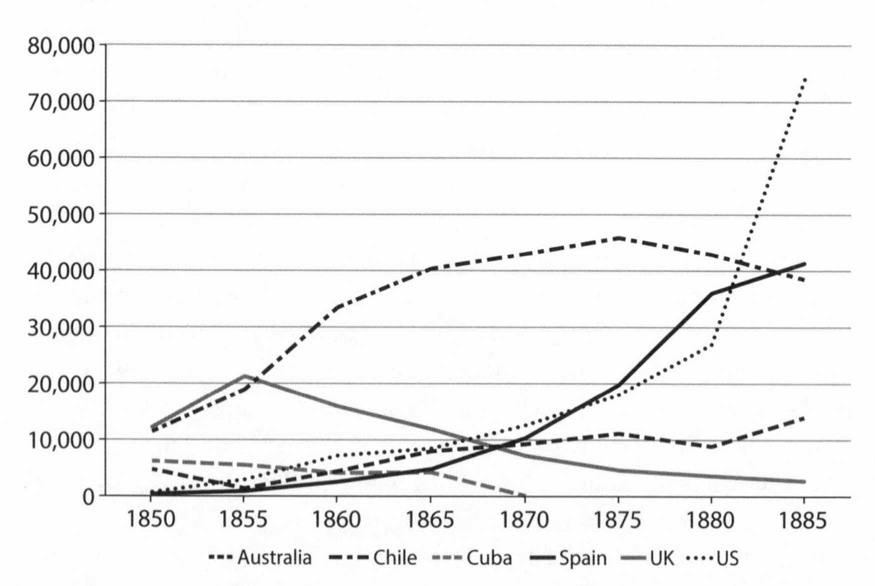

Figure 6.1. Copper production in selected countries, 1850–1885 (tons)
Adapted from C. J. Schmitz, *World Non-ferrous Metal Production and Prices, 1700–1976* (London: Frank Cass, 1979)

conductivity of their product and to master new technologies, especially in wire drawing. These were challenges that nimble-footed new businesses took to better than old firms, grounded in more traditional products and processes. They required investment in new equipment and the development of new skills in the labor force. Neither was easily achieved at Swansea, where capital was tied up in old plant, and employees in the sector were becoming more politicized. While Swansea firms studied the practices of their competitors and recognized that fundamental changes were needed, the obstacles they faced were daunting. By the early twentieth century, they were operating in an industry that had undergone a transformation in scale and structure, and despite undertaking some radical business reorganization, they could not match the size and capital resources of the big international companies now operating in the sector. The end of primary copper ore smelting in the Lower Swansea Valley in 1924 was an admission of defeat, bringing to an end a two-hundred-year-old industrial hegemony.

Crisis and Adaptation

The first real crisis period for Swansea Copper, which was acknowledged as such in the town, came in the late 1860s. The period of falling copper prices after the American Civil War exposed some harsh realities about the new world order in the industry. With the growth of smelting close to the sites of ore fields around the world, owners of copper ore mines had more choices at their disposal when it came to selling their product. By 1860 the practice of shipping parcels of ore to Swansea for sale at public ticketings was no longer the default option. Instead, companies increasingly looked to negotiate private sales with buyers, with the result that the ability of the Swansea smelters to work together to control prices was greatly reduced.[7] The Copper Trade Association, which had acted as the collective voice of the Swansea smelters since 1844, disintegrated in 1867 as individual firms abandoned their agreed quotas and prices in a bid to secure supplies of ore in an increasingly competitive market.[8] The effects of this breakdown in Swansea were manifested in the form of a shortage of ores and a loss of control over prices. This shift in the balance of power between ore sellers and the smelters of the Swansea District was articulated in 1878 by H. H. Vivian, head of Vivian & Sons, in a letter to J. M. Williams of the neighboring Swansea firm, Williams, Foster & Co. It advised that "it is our wisest policy to meet the wishes of the large sellers of ore so far as we can: these are not days in which we can dictate to them how they shall sell their ores."[9]

The implications of the changed trading conditions were also becoming apparent well beyond those directly involved in Swansea Copper. At a meeting of the town's harbor trustees in February 1868 it was noted that, even if trade soon picked up, "probably there would not be so much copper ore imported into Swansea as previously" and that the loss of revenue from fewer copper ore vessels entering the port was "a serious thing for the town."[10] Smelting companies tried to take measures to reduce their costs. In October 1870, for example, Williams, Foster & Co. petitioned their landlord, the Duke of Beaufort, for a rent reduction on the Rose and Landore copper works. Their case was not helped by the rising value of land in the Lower Swansea Valley, where demand for space for new industrial sites, notably the construction of the new Landore steelworks in the early 1870s, was pushing up prices. The reply they received was unsympathetic. The Duke's agent acknowledged the "change taking place in the copper smelting" but feared that "if he abated the Rent to [Williams, Foster & Co.] he would have applications from a score of other parties to make similar reductions."[11]

Faced with much less favorable market conditions in which to do business, Swansea's copper firms adapted their strategies, first by turning their attention to the smelting of other non-ferrous metals. George Grant Francis, former mayor of Swansea and author of the first history of the town's copper industry, noted in 1869 that "zinc is taking the place of copper here, and not only are large quantities made but some of the older copper works are now being converted into zinc manufactories."[12] Take-up of zinc (or spelter) production was widespread as demand for copper and zinc alloys rose. By the mid-nineteenth century, zinc was commonly used in a range of manufactured goods, such as water cisterns, pipework, and roofing, as well as in batteries and yellow metal, the copper-zinc alloy patented by George Frederick Muntz in 1832 for sheathing ships' hulls.[13] Vivian & Sons had been early proponents, commencing zinc smelting operations in 1835 to capitalize on the maritime demand for yellow metal as a ship sheathing material. Dillwyn and Co. were also early Swansea practitioners, smelting zinc at Llansamlet from around 1861.[14] But the key growth period for the Swansea zinc industry came in the crisis years for Swansea Copper in the late 1860s, when zinc smelting offered a lifeline to a number of Swansea firms encountering difficulties in the copper trade. Vivian & Sons opened a new zinc smelter at Upper Forest in 1868, and in the same year Pascoe Grenfell & Sons commenced zinc smelting operations at Upper Bank. At the Crown Copper Co., where copper smelting ceased in 1866, Williams, Foster & Co. installed plant for the smelting of zinc. New zinc companies were also attracted into the district to set up business, notably Villiers Spelter Co. Ltd. in 1873 and the Swansea Vale Co. Ltd. in 1876.[15] Although this marked a significant new concentration of zinc production the Swansea works relied on outside technologies, using furnaces developed in Belgium and Silesia and experts brought in from more established European zinc-producing regions to help in the early stages of operation. The shortcomings of this situation were exposed when it came to smelting the most complex ores such as the large zinc and lead deposits discovered at Broken Hill in South Australia in 1883. Swansea firms lacked the expertise to work the abundant new Australian deposits and were forced instead to purchase the zinc concentrates from Germany.[16] It was a telling example of how attempts at diversification exposed some of the limitations of Swansea's technological capabilities in non-ferrous refining.

As far as copper smelting was concerned, the Swansea firms focused increasingly on high-grade ores, which required multiple-stage smelting to remove impurities. These could still be smelted more profitably at Swansea using traditional furnace methods because they required more fuel-intensive treatment than low-grade oxides. Deposits containing a high metal content were discovered in South Africa and Newfoundland in the second half of the nineteenth century.[17] By 1880, the latter territory accounted for the greatest share of copper ore imports into the United Kingdom, at over 22,000 tons, with South Africa the second highest supplier, at almost 15,800.[18] Other specialist smelting functions that the Swansea firms sought to cultivate were the reduction of ores to extract precious metals. Soon after assuming control of Vivian & Sons, H. H. Vivian developed methods for extracting silver from argentiferous copper ores and mattes, drawing on the expertise of managers recruited from Germany.[19] In 1853, the firm of Lewis Llewelyn Dillwyn and Co. set up a silver smelting works at nearby Landore using ores imported from northern France.[20] Vivian & Sons and Williams, Foster & Co. took over the White Rock site for lead and silver smelting in 1870. Likewise, adaptations of the Welsh Process enabled the separation of gold from "white metal" during the process of obtaining Best Selected copper.[21] The specific knowledge and skills required to carry out these processes, and their reliance on fuel-heavy furnace treatment, meant that Swansea was able, at least for a time, to retain a foothold as a smelting center despite the general shift toward the consolidation of mining and smelting activities in ore-field locations.

While diversification of smelting activity became important in the difficult years of the late 1860s and 1870s, Swansea firms also placed increasing emphasis on manufactured output. To some extent this was nothing new. Since the 1730s, Swansea smelters had built mills equipped with water-powered hammers and later steam-powered rollers, in order to turn out a range of goods from battery ware to sheets and sheathing. By the 1860s, the future of yellow metal sheathing looked uncertain as a shift toward iron (instead of wood) in ship construction gathered momentum. Iron hulls were not susceptible to damage from seaweeds, worms, and barnacles, but were vulnerable to corrosion if they came into contact with copper. Both of these factors suggested that yellow metal's days were numbered, yet it was slow to fall out of favor. Copper alloys continued to be used to sheathe iron ships, with ship builders and repairers typically inserting an insulating layer of felt, rubber, or timber between the iron hull and the sheathing material to prevent a corrosive reaction between the metals. Despite contemporary criticisms of the cost and wisdom of this practice and the publication of alternative sheathing schemes,[22] demand for yellow metal sheathing remained high. It was a good thing for the Swansea copper firms that it did. The author of an article for the principal American trade journal, A. M. Levy, who had formerly worked as a chemist in a copper works in the Swansea District, described the production of Muntz metal sheets as "almost a necessary adjunct of the greater number of works," adding that "today the extraction of copper alone is an operation the profits of which are not sufficiently large to compensate for the capital that is invested."[23]

Yellow metal production at Vivian & Sons was concentrated at their site in Taibach, and rose from 120 tons in 1855 to almost 500 tons in 1892.[24] Much of the domestic share of this tonnage was in the form of sheathing and was sent to Liverpool and, to a lesser extent, the shipyards of the Clyde, where vessels were fitted with their protective cladding. In contrast, Pascoe Grenfell & Sons conducted their yellow metal business in partnership with G. F. Muntz, having become joint patentees when yellow metal was first launched in the 1830s.[25] Most of their yellow metal production was transferred from Upper Bank to the Muntz workshops in Birmingham in 1842.[26] This arrangement underlined the subordinate position of Swansea to Birmingham, which had five firms specializing in yellow metal manufacture by the 1860s, including Elliotts Patent Sheathing Company and the pioneering G. F. Muntz (figure 6.2). One mid-1860s account of the trade in Birmingham estimated that production of sheathing, ship bolts, nails, and wire made from Muntz metal amounted to eleven thousand tons per year, the value of which was an estimated £800,000.[27]

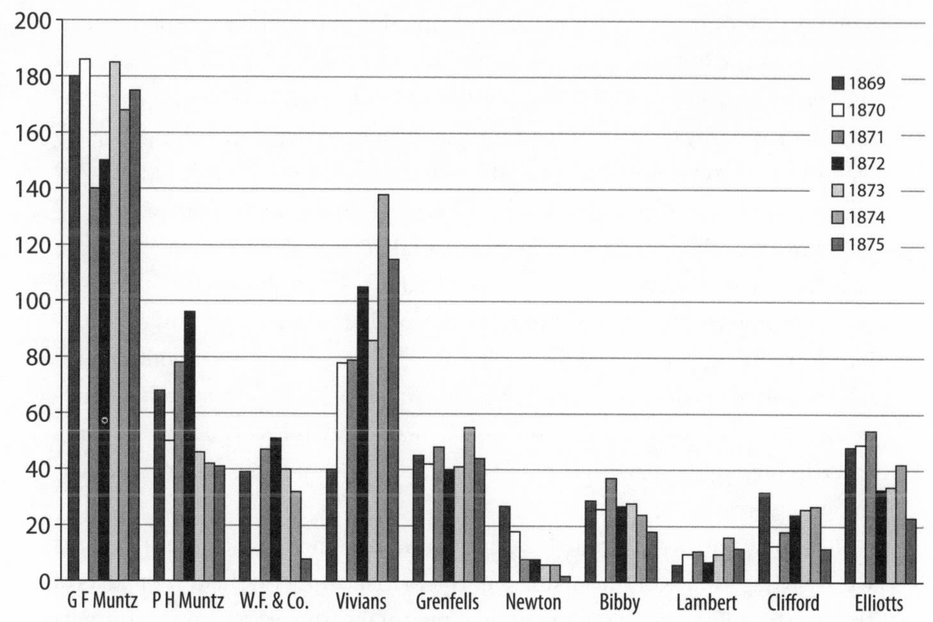

Figure 6.2. Number of vessels sheathed with metal supplied by principal firms, 1869–1875

Williams, Foster & Co. Memorandum Book, 12, 51–52. LAC/126/S1, RBA

Swansea firms faced competition from the Birmingham metal trades for other manufactured products too. The expansion of the railway industry and, in particular, the boom in the construction of steam locomotives by British railway companies in the middle decades of the nineteenth century, created significant new demand for copper and brass components. Some idea of the scale of the home market for locomotive parts can be gauged from a government audit of the railway industry, which found that, by December 1860, there were some 4,696 locomotives belonging to the railway companies of England and Wales and a further 781 in Scotland.[28] One contemporary estimate put the quantity of copper in a single locomotive at over ten tons.[29] Fireboxes, boiler stays, tubes, and plates made up the bulk of this tonnage and, with the trend toward greater engine power and enhanced speed, the size of these components was increasing.[30] The manufacture of boiler fireboxes involved the rolling of heavy copper plates, which were then shaped and finished. To some extent this kind of work was a natural progression for firms like Vivian & Sons and Williams, Foster & Co. that had a track record in the production of rolled goods such as sheathing, and both firms became trusted suppliers, working with the railway companies to establish accepted specifications as the industry became more standardized.[31] When it came to tubes, the processes were very different. Boilers of late nineteenth-century locomotives had as many as 150 copper and brass tubes passing through them, each weighing up to twenty-five pounds, which needed to be replaced at least every three years. They were made by drawing down thick cylinders of copper or brass to the required size using specialist tools. Tube making had been carried out in the brass mills of Birmingham since the early nineteenth century, and production continued to be dominated by Birmingham manufacturers like Charles Green and Thomas Bolton, who experimented with techniques to produce seamless tubes, making them safer for use at high temperatures.[32] By 1860 some 6,500 tons a year of seamless tubes were being made in Birmingham, with about five hundred men and boys engaged in their manufacture.[33]

It was not only the English Midlands where emerging clusters of firms were taking a share of the business of the British copper industry. In parts of the north of England where chemical manufacture was an important part of the economy, copper was also on the march, thanks to the presence of sulfur in a number of different forms of copper ore. As the residents of the smoke-choked Lower Swansea Valley knew to their cost, sulfur was a noxious by-product of copper smelting, but to chemical manufacturers it had an intrinsic value as a raw material. Sulfuric acid was an essential component in the manufacture of alkalis and fertilizers. Makers of soaps, bleaches, and other common chemical goods used it in large quantities and needed guaranteed supplies of sulfur for their production process. Traditionally British manufacturers had obtained their sulfur from mines in Sicily, but rising prices from the 1830s onward, along with successful experiments in the extraction of sulfur from cupriferous pyrite, led to the adoption of pyrites as their primary source of this key raw material. Found in abundance in southern Spain and Portugal, pyrites was first imported into Britain by chemical makers like John Allen of County Durham, who developed methods of burning it to extract the sulfur content.[34] British imports of copper and iron pyrites increased more than fourfold in less than a decade, from 93,889 tons in 1861 to 411,512 tons in 1870.[35]

This development had significant implications for the copper industry. Large residues of metal oxides left after the extraction of the sulfur were of little use in the alkali factories, but a "wet process" to separate the copper from this metalliferous residue was patented by Glasgow chemist William

Henderson in the 1860s.[36] The development greatly enhanced the attractiveness of pyrites to metal producers as well as to chemical manufacturers and led to a growth of copper production in districts where alkali manufacture was already established, especially in Lancashire and, to a lesser extent, the northeast of England. Copper works had long been part of the economic landscape of the northwest, where ores from Anglesey and later from Chile, via the port of Liverpool, had been smelted at sites in Ravenhead and St. Helens, but the opportunity to purchase burnt pyrites from the alkali factories of the district, and process it using the Henderson method, gave them a new lease on life. William Keates's Sutton Copper Company in St. Helens, and the Bridgewater Smelting Company, established by John Knowles Leathers and John Wilson, built alliances with local chemical businesses to this end.[37] The Liverpool-based chemical industrialist, Charles Tennant, moved in the opposite direction, capitalizing on the new methods of refining pyrites by extending his family interests in the alkali trade into mining and metal refining. On his initiative, the Tharsis Sulphur and Copper Company was incorporated in 1866. It began working the pyrites deposits in the Huelva region of Spain, leading to an uplift in mining activity there in the 1860s and 1870s. The venture secured an abundant supply of sulfur for its chemical businesses and, by 1872, was supplemented by the acquisition of seven metal-extraction businesses in Britain, including one in Wales, at East Moors in Cardiff, as outlets for its substantial metal residues.[38]

The effects on the locational diversification of the copper industry in Britain were lasting. The industry had never been confined solely to Swansea, but it was now more geographically spread than ever. Of the twenty British-based firms processing raw copper at the end of the nineteenth century, only eight were situated in the Swansea District. The rest were located in the chemical-industry heartlands of Lancashire and the Tyne, as well as the Midlands metal-manufacturing region (table 6.1). Swansea's claim to be the home of the British copper industry was weakening.

A New World Order

Despite the erosion of its leading position in the industry, the Swansea District was still a commercially viable presence in the world copper trade in the early 1880s. Those firms that had weathered the storm of the post–American Civil War years had done so through their responsiveness to shifting patterns of international trade and an ability to increase their manufacturing output, but this did not necessarily equip them to withstand an even greater wave of

TABLE 6.1
British manufacturers of "Tough" and "Best Selected" copper, 1897

Company	Location
Bede Metal & Chemical Co. Ltd.	Tyneside
John Bibby, Sons & Co.	Merseyside and Lancashire
Thomas Bolton & Sons	Staffordshire and Lancashire
Broughton Copper Co. Ltd.	Lancashire
Cape Copper Co. Ltd.	Swansea District (Briton Ferry)
Elliott's Metal Co. Ltd.	Birmingham
Henry Hills & Sons	Anglesey
Charles Lambert & Co.	Swansea District (Port Tennant)
Landore Copper Co.	Swansea District (Lower Swansea Valley)
Logan & Co.	unidentified
Nevill, Druce & Co.	Swansea District (Llanelli)
Newton, Keates & Bolton	Lancashire
Rio Tinto Co. Ltd.	Swansea District (Cwmavon)
W. Roberts	unidentified
St. Helens Copper Co. Ltd.	Lancashire
Tharsis Sulphur & Copper Co. Ltd.	Lancashire and Cardiff
United Alkali Co. Ltd.	Lancashire
Vivian & Sons	Swansea District (Lower Swansea Valley)
H. H. Vivian & Co. Ltd.	Swansea District (Lower Swansea Valley)
Williams Foster & Co. and Pascoe Grenfell & Sons Ltd.	Swansea District (Lower Swansea Valley)

Source: Copper Manual: Copper Mines, Copper Statistics and a Summary of Information on Copper (New York: D. Houston and Co., 1897), 139.

development that commenced with a new phase of expansion in the American copper ore mining industry in the late 1880s. The opening up of vast new ore fields in Montana and Arizona yielded sufficient new supplies of copper to satisfy not only domestic American smelting demands but also a surplus for export.[39] The copper output of the United States rose at a staggering rate, from just over 74,000 tons in 1885 to over 270,000 tons in 1900. In a decade it had mushroomed again to 485,804 tons in 1910.[40] The increase was driven by the cutthroat business approaches of the Anaconda Company and its rivals. The aim of these firms was to maximize their scale of operations and flood the market with low-price copper in an attempt to force their competitors out of business. To do so, they integrated vertically and horizontally to achieve maximum economies of scale, embracing not just mining and processing sites but a host of ancillary businesses such as brickworks, sawmills, electrical plants, and foundries. These mighty new "industrial metabolisms" concentrated huge resources on the mining and processing of low-grade copper ore for maximum profit and did so on a scale unmatched by anything the industry had seen before.[41] This new phase propelled the United States into a position of world dominance in copper production, notwithstanding the rapid advance of new producing nations like Japan and Mexico (figure 6.3) and heralded a new era in the global copper trade.

The newer companies active in the world copper industry by the latter decades of the nineteenth century commanded much larger reserves of capital and influence over mining, smelting, and processing operations. The Anaconda Company had capital assets of $52.5 million by 1898.[42] In contrast, the three largest Swansea copper companies active in the period—Vivian & Sons, Williams, Foster & Co., and Pascoe Grenfell & Sons—were family firms, each headed by a descendant of their original founder. Their capital assets, though substantial for the region in which they operated, were not large by global standards. Vivian & Sons, for example, had capital of £1,123,000 by the end of the 1870s.[43] There were a few firms operating in the Swansea region that resembled more closely their new international rivals. The London-based Cape Copper Company, which owned copper mines, a smelter, and railway in Namaqualand, as well as the Briton Ferry copper works in the Swansea District, became a limited company in 1888 with capital of $600,000.[44] Likewise the Cwmavon copper works was taken over by the multinational Rio Tinto Company in 1884. Like the Tharsis Sulphur and Copper Company,

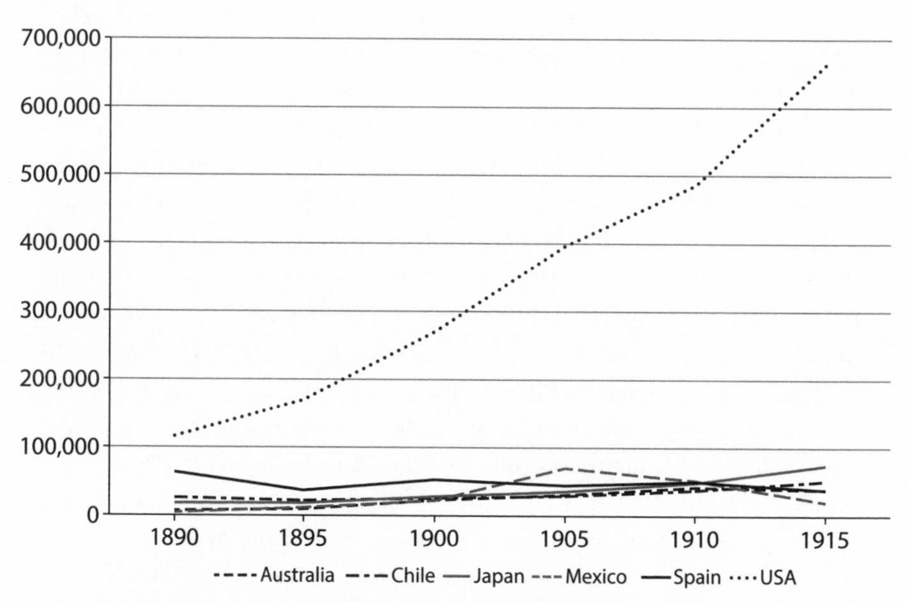

Figure 6.3. Copper production in selected countries, 1890–1915 (tons)
Adapted from C. J. Schmitz, *World Non-ferrous Metal Production and Prices, 1700–1976* (London: Frank Cass, 1979)

Rio Tinto had extensive mining interests in the low-grade ore fields of southern Europe, and the two companies' combined assets amounted to some $22 million by the late 1880s.[45] The reach of the German commercial grouping known as Metallgesellschaft, probably the most powerful presence in the global nonferrous trade in the late nineteenth century, also extended to Swansea, where it was a shareholder in Williams, Foster & Co. and the Swansea Vale Spelter Works.[46] Through the shrewd marketing skills of its founder, Wilhelm Merton, this company had built up a dense network of investments in metal refineries and fabricators.[47] In comparison, the business structures of the older Swansea businesses looked old fashioned. Pascoe Grenfell & Sons did not become a limited liability company until 1890. Vivian & Sons waited even longer, until 1914. By the 1880s their traditional operating principles were increasingly out of step with global trends, and their ability to exert influence in the industry was diminishing.

One consequence of the smaller size and capital resources of the Swansea firms was that they lacked the reserves to undertake ambitious levels of investment in new technologies. By the late 1880s, rapid advances were being made in methods of processing copper ore. Bessemer converters, commonly used in the production of steel from molten iron, were adapted for use with copper following experiments in England and France in the early 1880s. They allowed the cheap and efficient production of "blister copper," ready for refining, and were installed at the mines in Butte, Montana, from 1885 onward. The technology was well known to Swansea firms. Vivian & Sons began experimenting with bessemerization in 1885 and soon introduced the process in their silver plant at White Rock,[48] but as one contemporary British metallurgist noted, the method was only really cost effective if implemented on a large scale, "for continuous operation of the whole smelting plant."[49] In the US industry, take-up of the new process was much more ambitious. There, Bessemer converters were used to increase the volume and rate of the smelting process. One installed at Bisbee, Arizona, in 1894 was capable of handling six to eight tons of regulus and producing blister copper of 96–97 percent purity. This was some three times greater than the rate at which regulus could be refined at Swansea. The town's newspaper, *The Cambrian*, noted with some irony that it was a former Swansea man, Henry W. Edwards, who had installed the new plant at Bisbee. The role of Welsh experts in the advance of the copper industry in the United States was a source of both pride and frustration in Swansea: "A few such improvements are badly wanted in our Swansea copper works, the owners of which are being left behind in these days of severe competition and progress."[50]

As the case of Bessemer technology squarely demonstrated, it was not so much in their knowledge of new techniques or in their desire to innovate that the Swansea copper firms lost out to their American counterparts, but rather in their ability to invest in wholesale implementation. With much of their capital tied up in plant devoted to coal-fired furnace smelting, and located in a region where coal was cheap and readily accessible, Swansea firms did not find the case for investing in new, coal-saving techniques compelling. The advent of electrolytic refining methods to process blister copper brought this issue into even sharper focus. Electrolytic refining had been patented by James Elkington in 1865 and was in use at his copperworks in Pembrey, on the western fringe of the Swansea District, by the late 1860s.[51] It yielded copper with a very high metal content and excellent conductivity properties, making it particularly suitable for electrical applications. The spread of electrical technologies in the nineteenth century offered new market opportunities for copper manufacturers. It took several decades for those opportunities to acquire critical mass and, when they did, it was American producers rather than the long-established Swansea firms that were better placed to capitalize.

Electrical telegraphy swept across Europe and North America in the 1840s and 1850s. Initially it made only modest demands on copper manufacturers.[52] Although a highly conductive metal—in its pure form it is matched only by silver—copper wire was eschewed by the pioneer telegraphy companies. There were good reasons for this. Copper lacked the tensile strength needed to provide suitable material for overhead telegraph wire, and its status as high-value metal made it vulnerable to theft and thereby "inapplicable for open air lines."[53] Iron wire was almost universally preferred. Although far less conductive, it was mechanically stronger and less attractive to thieves. The chief effect of early telegraphy was not, therefore, to extend the market for copper but to stimulate research into copper's chemical and electrical properties. The leading specialist in the field, Augustus Matthiessen, who carried out critical work in the 1850s and 1860s, established that small variations in the chemical composition of copper could have profound effects on electrical performance.[54] On a scale of 0 to 100, where 100 represents maximum conductivity, Matthiessen rated Rio Tinto copper at just 14.24. Metal smelted from Burra ores, on the other hand, scored 88.86, while Lake Superior copper led the field with 92.57.[55] Identifying highly conductive brands of copper was a key consideration for the makers of submarine telegraph cables. The enormous length of undersea cables made them highly resistant to the flow of electrical current. Iron wire was of no use here; copper had to be employed if the resistance was to be overcome. And copper could be used because of the protective armoring applied to cables to overcome copper wire's tensile deficiencies, while the great depths at which the wire was laid offered security against theft.

Submarine telegraphy remained an experimental field in the 1850s, yet it was a field in which Swansea copper companies were well placed to intervene. It was a British-dominated business, closely aligned with British imperial interests, commercial and strategic, with obvious opportunities for British copper manufacturers.[56] Sure enough, the early submarine cables were made with Swansea copper. The first, abortive Atlantic cable (1857) used wire drawn out by Thomas Bolton & Co. at their Oakamoor wire works in Staffordshire, using metal supplied by Williams, Foster & Co. of Morfa. Thomas Bolton & Co., a key player in the nascent industry, also worked in collaboration with Sims, Willyams, Nevill, Druce & Co. of the Llanelly Copper Works.[57] The success of the first fully operational transatlantic cable in 1866 set off a boom in cable laying in the 1870s and 1880s. Even so, the volume of copper embodied in cables was relatively slight. The two high-speed transatlantic cables laid in 1894 revealed just how little. That of the Commercial Cable Company extended for 2,161 nautical miles, with a total weight of 5,460 tons, but the cable's insulation and armoring accounted for almost all of that weight. The conductive core took up just 495 tons of copper, equivalent to 513 pounds per nautical mile. The cable of the rival Anglo-American Telegraph Company was a little weightier but not significantly so, with a copper core of 650 pounds per nautical mile.[58]

By 1894, 152,000 miles of submarine cable had been laid down worldwide—one of the great technological accomplishments of the age. The undersea network was dwarfed, nevertheless, by landline cabling, which extended to around 2 million miles.[59] In this, copper was beginning to feature more prominently. "Hard-drawn" copper wire, which was mechanically robust, was "coming into use rapidly" in the 1880s according to one of the leading US suppliers, John A. Roebling's Sons Co. of New Jersey.[60] Iron wire, with a conductivity of 10 (on a scale of 0 to 100), would never do for high-speed telegraphy and telephony, sectors that were experiencing rapid growth.[61] Hard-drawn copper, on the other hand, with a conductivity of 94, was well adapted for those purposes. However, because the "process of hardening reduces the electrical conductivity of the metal some 2 to 4 per cent," only copper of superlative character was eligible to be "hard drawn."[62] That tended to exclude Welsh copper. On the London market, it was copper imported from the United States, the native copper of Lake Superior especially, and Japan that was "in great request for the manufacture of wire for electrical purposes."[63] Even this had to be electrolytically treated before being subjected to the hardening process. The remaining Swansea firms were equipped to do so. Vivian & Sons had established an electrolytic refinery at Hafod, and it was noted in 1898 that "very large quantities of the Lake Superior copper are dealt with" at the Hafod works,[64] but with implementation possible wherever electricity was available, there was no locational advantage at Swansea. It was noted in one study that "the process of electrolytic refining is of such a nature that it cannot well be connected with an ordinary smelter. It requires cheap power, highly trained chemists and electricians, and a general scale of operations too large to be in harmony with any but an exceptional smelting plant."[65] These were conditions best met in Montana or Arizona, not in South Wales. The giant concerns of the American West, where hydroelectric power was abundant, were far better suited to take-up of electrolytic processing. The mining center of Butte, Montana, had ready access to hydroelectric power, making the adoption of electrolytic processing highly cost effective.[66] There was also more incentive for them to do so. The electrical industries accounted for over half of all copper consumed in the United States by the early 1920s, with the production of electrical goods, power lines, telephone, and telegraph cables all accounting for a growing share of the market.[67] In comparison, progress in Britain was slow. Legislative restrictions hindered investment in the electrical manufacturing sector, and demand in some sectors, such as domestic electrical goods, was small compared to other countries, especially before 1914.[68]

These diverging circumstances set the US and British copper industries on different trajectories. The rampant American industry was mining and smelting more than enough copper to meet the needs of its domestic markets, and was capable of refining to a high enough purity to supply the growing electrical sector. The output of the Swansea firms remained focused on standard copper, of at least 96 percent metal content, which was more suited to the manufacture of alloys and for engineering and mechanical applications.[69] Imperial markets became an increasingly important source of orders for these products, with India the most important of these.

Copper exports to India from Britain had been controlled by the East India Company since the eighteenth century, but following the demise of the company in 1858 the Swansea firms continued to serve Indian customers with manufactured goods and semi-products, both directly and indirectly. Demand for the locomotive parts turned out by Swansea firms, for example, was buoyed by the dramatic expansion of the Indian railway network in the second half of the nineteenth century. With over 15,600 miles of track laid by 1890, but a relatively small domestic locomotive-building sector, India looked to British firms to supply rolling stock. Some twelve thousand locomotives were exported to India between 1865 and 1941.[70] The scale of this market, added to home demand, placed the British locomotive-building industry in "a state of great activity" in the early twentieth century,[71] with knock-on effects for Swansea's copper firms, positioned further down the supply chain. Copper goods also featured prominently in the so-called "bazaar trade," through which dealers located in India's major ports and distribution centers imported and sold a whole variety of items in demand by the general population. Yellow metal, sheet brass, and wire made by firms in Swansea, Birmingham, and elsewhere accounted for some 15,764 tons of goods exported from Britain to India in 1864.[72] As a share of the value of Indian imports, metals grew from 7.5 percent in 1880–1881, to 11.2 percent in 1910–1911.[73] Copper squares for the Indian market were a specialty of the Swansea District. These items contributed to the estimated thirty thousand tons of copper that was consumed in India annually by the late 1890s, mainly in the production of copper and brass utensils and cooking pots.[74] It was an important source of custom that the Swansea firms worked hard to satisfy. In 1882 John Cady, of Williams & Foster's Morfa works, was required to send instructions on how best to restore to mint condition yellow-metal sheets that had discolored in transit to India. Rubbing them with emery paper or re-annealing them before rubbing them by hand with hemp was the recommended method.[75] Such customer interaction was vital to retain a share of the Indian market, where low prices and exacting standards regarding the appearance of goods were paramount.[76]

Egypt and the near East was another important market for manufactured Swansea copper goods. The supply of copper bottoms for Egypt, in particular, was dominated by Swansea firms in the late nineteenth and early twentieth century, despite the market being "a most difficult one, calling for a very high standard of quality at a very low price."[77] The demand was very likely caused by the expansion of the Egyptian sugar refining industry from the 1890s, creating a market for the kind of specialist copper bottoms that Swansea firms had been supplying since the eighteenth century for manufacture into boiling pans and other refining vessels for the processing of sugar cane in the Caribbean and to the brewing industry. Traditionally, Egyptian sugar cane had been exported for processing, but the rising price of sugar in the 1890s led to the establishment of seven new companies to refine the crop internally.[78] It was a trade in which the Swansea firms faced little competition, but the same could not be said for the supply of other specialist components for export. By the early twentieth century, a number of rival European producers, especially in Germany, had achieved success in meeting the demand for products destined for the same markets as the Swansea firms. India sheets and squares were being produced at Osnabruck. At Mansfeld Copper works some four hundred tons of fireboxes were produced per month using a hydraulic pressing method rather than traditional hammering techniques; while at Nestersitz Poemmerle, near Aussi, one hundred tons of heavy copper firebox plates were produced per month.[79]

Nevertheless, foreign orders for specialist manufactured products, to some extent at least, helped Swansea copper firms to avoid the uncomfortable reality that they had made no significant headway in the real growth area of the industry: namely the supply of copper for electrical goods. Although there were individuals in Swansea who were aware that the town's copper trade was being left behind, the public face of the town rarely reflected concern for the future. In other spheres of trade there was cause for optimism. The growth of the tinplate, chemicals, timber, and patent fuel sectors helped mask the difficulties in the copper trade.[80] The opening of a new dock in 1881, marked by a royal visit, was both the cause and the effect of increasing volumes of shipping using the port (table 6.2). Some commentators

TABLE 6.2
Port of Swansea: total imports and exports in selected years, 1875–1915 (tons)

	Imports	Exports
1875	518,000	1,031,000
1880	648,799	1,333,093
1885	692,372	1,792,332
1890	697,643	2,271,603
1895	623,183	2,388,465
1900	845,110	3,259,404
1905	873,167	3,652,889
1910	993,952	4,789,104
1915	809,003	5,067,474

Source: W. H. Jones, *History of the Port of Swansea* (Carmarthen: Spurrell, 1922), 365.

reveled in the growing diversity of the regional economy and viewed it as a sign of Swansea's trading strength rather than a symptom of the weakening of its staple trade.

Buoyant levels of employment in the copper industry itself also helped to conceal its long-term decline. The increase in output of manufactured products by Swansea copper firms, and the labor intensity of the manufacturing operations, changed the employment structure of the industry locally. The typically compact smelting workforce described by Frédéric Le Play in the 1840s was swelled by the addition of operatives responsible for different parts of the manufacturing process. At Morfa in 1873, the yellow-metal mill employed 39 workers, including adults and children, at tasks ranging from the heating and rolling of the sheets, to the cutting, shearing, and washing of them, and the weighing and warehousing of the finished article.[81] The Morfa copper mill, where items such as the plates for locomotive fireboxes were made, was a much larger operation, with over 150 workers. Here, employees were engaged in preparatory tasks required to make the copper pliable, including furnace treatment, rolling, cooling, and pickling of the metal. Then there were a range of jobs specific to the production of particular manufactured items, including cutting, shearing, "splatching," and hammering to achieve the shapes and dimensions required by customers. The operation of mills also necessitated the employment of a team of mechanics and tradesmen to keep the machines and tools used by the operatives in good working order. At Morfa there were fitters, smiths, carpenters, masons, patternmakers, enginemen, and firemen on the payroll.[82] Some of these employees may not have been "copper workers," in that their labor was not connected directly to the process of producing copper, but their employment at firms like that of Williams, Foster & Co. helped bolster the total number of workers employed in copper in South Wales. Taken as a whole, this was higher in 1901, at 2,386, than it had been in 1851 when the total number of employees was 2,148.[83]

The healthy employment levels in the industry go some way to explaining why, when the first of the big-name copper firms, Pascoe Grenfell & Sons, went into liquidation in 1892, the reaction in the town was one of shock. *The Cambrian* newspaper reported the firm's announcement with an air of disbelief: "To say that the news was startling is to speak in the mildest terms. Everybody in the town, including the whole of the most trusted employees of the great firm of Messrs. Pascoe Grenfell & Sons, had come to consider the firm as one of the soundest and most lasting commercial organizations in the district."[84] In fact, the firm had been attempting to limit its losses for some time. After reorganizing itself as a limited liability company in 1890 it set about reviewing its operations in Swansea. The decision was made in February 1891 to end sulfate production, and just seven months later the board resolved to close the yellow-metal mill at Upper Bank. When the final resolution was passed to wind up the remaining operations it was noted that "the company cannot by reason of its liabilities continue its business."[85] The blow was softened by the takeover of their remaining operations by Williams, Foster & Co., who continued to run Middle Bank and Upper Bank works,[86] but the end of the Grenfell era was a powerful symbol of the changing fortunes of Swansea Copper.

Management and Labor

The Cambrian, rather naively, surmised that the liquidation had come about "as much from family and personal considerations as from the recent unprofitable condition of the copper trade."[87] This was an indirect acknowledgment that times had changed since the passing of some of the dominant figures in Swansea's Victorian copper industry, notably Pascoe St. Leger Grenfell, who died in 1879. There was a changing of the guard in some of the other Swansea firms too. H. H. Vivian died in 1894, having guided the fortunes of Vivian & Sons for forty years. His younger brother, A. P. Vivian, who had overseen the firm's copper works at Taibach, to the east of Swansea, while H. H. Vivian's attention focused on Hafod, remained actively involved after his brother's death, but extensive political commitments and a main residence in Cornwall meant that he was often absent and did not scrutinize business affairs as closely.[88] Yet this did not necessarily mean that effective business leadership was lacking. By the end of the nineteenth century the surviving Swansea firms were relying more and more on the oversight of managers to run their operations. While managers were responsible for carrying out the decisions of the firms' partners and communicating with them about the running of the works, there was also scope for innovation and individual initiative in these posts. Even if they had no direct access to capital resources or strategic planning decisions, talented managers could increase profit margins and enhance output with incremental improvements to technical processes and workplace organization. The work of Thomas Nicholls, assistant manager at the Cape Copper Works in Briton Ferry, showed how.

Nicholls took up the post of assistant manager at Briton Ferry in 1889, after a lengthy career in the service of Williams, Foster & Co. at Morfa. His deep technical knowledge of smelting and refining processes and an interest in modern management practices led to the introduction of a number of improvements at the works under his watch. His best-known invention was the "direct refinery process," jointly patented in the early 1890s with his associate, Christopher James. This was a modification of the traditional Welsh Process, which bypassed the roasting stage in the conversion of white metal into refined copper. The purpose, Thomas Nicholls explained, "is to do away with the expensive and wasteful operation of roasting, in which a large proportion of the copper is returned to the work as roaster slags and a lot of copper always locked up in the roaster bottoms and to work with the cheaper calcination." Later he proposed a modification to the process of bessemerizing copper matte, removing the need for a clay lining in the converter, which required replacing after every five charges. Ever the pragmatist, Nicholls emphasized that the important feature of these improvements was that they could be adopted without need of expenditure on new equipment or staff.[89] In a region wedded to traditional coal-fired furnace technology, time- and cost-saving adaptations of existing processes were the favored methods of modernization.

Working practices in the labor force (figure 6.4) at the Cape Copper Company also came in for close scrutiny. Nicholls, who may well have encountered new methods of scientific management when visiting the United States, made further attempts to improve efficiency by attempting to regulate worker behavior and enforce strict habits of timekeeping and productivity. He set out clearly defined tasks for different categories of worker. In the cupola department, for example, smelters were required to attend to the working of the furnace below the charging floor, help make wells, attend to sand moulds, and patch the furnace when required. The feeders' role was to charge the furnaces, assist in tapping and patching, and cut down lumps in the furnaces. Helpers had to attend to the slag trams, fetch clay, coke, and other materials, assist in patching and tapping, and deliver all metals and regulus to the cupola house. Firemen were required to remove slag and dispose of it safely. A series of rules and fines were imposed to ensure prompt attendance and diligence while at work. Checks (workers' identity tokens) were to be taken up promptly at the start of a shift (either 6:30 a.m. or 5:30 p.m.) with a one-shilling fine for every half-hour a worker was late. Failure to ensure cleanliness of workspaces, and instances of disorderly conduct, drunkenness, and swearing were all subject to five-shilling fines, while cases of theft or damage to works property were liable to dismissal and legal prosecution.[90]

Figure 6.4. Workers at Cape Copper Company Works, Briton Ferry, c. 1890. T. D. Nicholls collection, WGAS

These kinds of measures were designed to reduce loss of worktime and minimize absenteeism, especially at times of the year when local distractions such as the annual Neath Fair tempted men away from the workplace. Yet the social climate of late nineteenth-century Swansea was not an easy one in which to attempt to enforce greater levels of control over working patterns. The growth of labor politics and new unionism in late nineteenth-century Britain instilled in industrial workers a sense that their labor was a valuable asset and increased levels of labor organization. Copper workers were by no means in the vanguard of this movement. Along with speltermen, they were identified at the annual Trades Union Congress in 1887 as among "the toiling thousands [who] neglect to combine for their collective well-being."[91] Where grievances arose their first route to redress was often via their works managers. At Morfa, in May 1873, against a backdrop of rising provision prices, Morfa millmen wrote to their manager, Mr. Martin, to appeal for a 35 percent increase in their wages, as "we find it impossible to maintain ourselves and families at the high price of the necessaries of life and we consider at the same time our labour fully worth the said advance."[92] Their appeal received short shrift. A number of workers who made the decision to join the ranks of a local union, possibly the fledgling Association of Tinplate Makers, established in 1871,[93] were unceremoniously locked out, and left to depend on payouts from a support fund set up by sympathetic fellow industrial workers in the district.[94]

Resistance toward the influence of trade unions reflected the strong traditions of paternalism in the Swansea copper industry. Liberal industrialists like H. H. Vivian and his counterparts in the other Swansea copper works viewed the influence of trade union representatives over their employees as unwelcome outside interference, but even without the formation of a specific copperworkers' association, there was no shortage of trade union leadership available to represent them. The local branch of the dock workers' union, in particular, actively recruited members from the metal industries.[95] During a dockworkers' strike in the early summer of 1890, the General Secretary of the London Wharf and Dock Labourers' Union, Tom Mann, addressed a mass meeting of workmen in Swansea's Albert Hall in July 1890 and threw out a direct challenge to the copper workers present:

> There are something like 2,000 men employed at the Copper Works, and I say that in my opinion you are most scandalously paid. (Applause) Hark you, I think the blame rests on your own shoulders. Therefore, do not blame your employers. The tin men get eleven or twelve bob a day for 8 ½ hours. That is worth looking at. What does the rollerman get? Five shillings a day! (A voice: "Half of that.") . . . Now what is the difference? Why one is organised and the other is not. . . . There is nothing to prevent you coppermen, especially in the mills, to get double the wages you are getting now.[96]

In practice, there were significant variations in rates of pay in the copper industry, depending on the type of work undertaken. At Morfa, all the employees in the yellow-metal mill (except the foreman) earned less than six shillings each per day, with child workers paid less than two shillings. Even after a slight wage increase awarded in April 1873, this amounted to a relatively modest daily wage bill for Williams, Foster & Co., for this part of their operation, of around £7. In the Morfa copper mill, the highest paid employee was the head refiner, who took home nine shillings and sixpence per day, while most of the others, similar to their counterparts in the yellow metal mill, earned a daily rate of less than six shillings. While payment by day rate was the norm in the mills, piecework was more common among furnace workers in the Swansea copper works, as in many other industrial sectors in the late nineteenth century. For copper smelters, this meant payment by ton of furnace output. In the Cape Copper Works in 1890, for example, roaster men were paid three shillings and six pence per ton of blister tapped, with three tons in twenty-four hours being the usual rate of production. While these piecework rates may have delivered a living wage for workers during normal periods of production, the unpredictability of working with variable raw materials meant that there was always a degree of uncertainty about output levels, and therefore earnings. Moreover, in circumstances such as shortages of ores or coal, or downtime for furnace repairs, companies reserved the right to transfer piece workers onto day laboring rates until regular production could be resumed.[97] It was a situation that undermined workers' sense of financial security and became a bone of contention in the increasingly adversarial environment of the late nineteenth-century industrial workplace.

At Hafod in 1904, cupola furnace workers who were paid a piece rate struck over the management's refusal to guarantee them a day rate of five shillings. The local press was sympathetic, noting that their earnings had fallen from five shillings and six pence "a turn" to about four shillings "owing to altered conditions over which they have no control."[98] At the Cape Copper Works in nearby Briton Ferry in 1905, refiners and furnacemen complained that their rate of pay was not sufficient for them to employ a full "gang," including helpers, on a par with the practice in other works. Again, managers gave no ground, refusing to consent to arbitration and instigating a lockout.[99] The first decade of the twentieth century was peppered with further stoppages as furnace workers and mill men sought redress for what they saw as disparities in pay and conditions. Recurring grievances included not only dissatisfaction with pay but also issues such as Sunday working hours, holiday entitlements, and lengthy shift patterns.[100] Far from viewing the diminished status of Swansea firms in the global copper trade as a disincentive to strike, or as a threat to their livelihoods, these workers saw it as strengthening their claim for better pay and conditions. One striking worker at Morfa in 1890 identified the redress of workplace grievances as essential to prevent the loss of skilled labor to rival overseas firms: "Our best and most skilled men," he observed, "leave this country for foreign lands, where they are paid double the wages we are . . . for the simple reason that they are not being properly treated and are anxiously sought for in foreign lands."[101]

The First World War and Beyond

Some of the deepening problems faced by Swansea Copper were temporarily suspended during the war years of 1914–1918. At first there were fears of hardship in the industry due to disruptions in trade. The Cape Copper Works in Briton Ferry endured "three months of great slackness" in the early stages of the war owing to the non-arrival of copper ore from its mines.[102] At Vivian & Sons copper works in Taibach, a Mutual Help Society was formed by the workers to provide relief to those who might lose their jobs owing to war conditions, but these fears proved unfounded, and the Society was able to donate its funds to charitable causes instead.[103] It was the formation of the Ministry of Munitions in 1915 that made the difference. The use of copper for purposes other than munitions was prohibited and, under the new ministry's oversight, production of mortars, explosives, guns, tanks, and artillery soared. Output of shells, for example, rose from 7.4 million in 1915 to 51.6 million in 1916 and 87.7 million a year later.[104] It was a similar story in the United States, where huge wartime demand for copper saw the industry expand with new mines opening in Arizona, New Mexico, and Utah, and rapid increases in the capacity of electrolytic refineries.[105] High demand, along with high copper prices, enabled the firms to make healthy profits in the 1914–1918 period. Rio Tinto, at its annual meeting in London in 1917, reported to its shareholders that "large profits" had been secured, despite difficult operating conditions and some additional costs.[106]

Although there were profits to be made from the unusual wartime demands for copper, it was the question of what would happen after the end of the conflict that began to loom large in industry and government circles by 1918. In both Britain and the United States, wartime legislation to prohibit trading with enemy nations had undermined the dominance that German marketing firms had exercised over copper purchasing since the latter years of the nineteenth century.[107] Both countries also resolved to implement measures to prevent such dominance from occurring again. In Britain, the Non-Ferrous Metals Trade Committee of the Board of Trade, aware of the ground that had been lost before the war, recommended that the government should provide financial assistance for the development of copper refining in the United Kingdom, and emphasized the "great importance" of preventing German domination of the industry in the future.[108] The result of these recommendations was the formation of the British Metal Corporation (BMC) in 1918, with the aim of continuing state controls.[109] Firms wishing to trade in nonferrous metals were required to obtain a license from the Board of Trade—a measure introduced specifically to counter the threat of renewed German involvement in the industry. But it was also hoped that the BMC would answer some of the problems experienced by Britain's comparatively small firms in being unable to wield much influence in the industry.[110] In the United States, similar discussions were taking place. A Copper Exports Association was formed, representing some 75 percent of American copper producers. Both the British and US initiatives demonstrated a determination to avoid the circumstances that had led to the dominant position of Metallgesellschaft, but also a realization that this kind of collective organization was necessary in order to exert influence in the global copper industry.

Despite the attempt to put measures in place to protect the industry after the war, falling prices and a reduction in demand for copper produced difficult market conditions. Levels of copper consumption in Europe lagged well behind prewar levels, and with large stocks of copper remaining from the end of the war there was little capacity to absorb new production.[111] The Cape Copper Company was a casualty, with the closure of its mines in 1919.[112] What really pushed Swansea's remaining copper firms to contemplate fundamental reorganization was the collapse in prices in the global copper market in 1920. Prices fell as demand for copper in what had been key consuming countries, like Japan, began to falter. By March 1921, copper prices had fallen by 35 percent since the previous August. Producers were forced to restrict output, and unemployment rose in copper-reliant manufacturing sectors like the brass and electric industries.[113] The effect of these conditions was made even worse by the collapse of some continental currencies in the early 1920s, which dealt a blow not only to European trade but also to the market for copper goods in the East. For Williams, Foster & Co., it made "the reorganisation of works and modernisation of plant . . . a necessity."[114] Old firms, already struggling to compete with leaner, newer rivals, had little resilience in the face of economic downturn. Vivian & Sons were in the same predicament as their neighbors, Williams, Foster & Co., and came close to going out of business in the early 1920s. As a solution to their shared problems, the two firms agreed to undergo a merger.

The British Copper Manufacturers Ltd. was incorporated on November 17, 1924. It consisted of the business assets of Williams, Foster & Co. at Morfa and Middle Bank, and Vivian & Sons at Hafod and Taibach, as well as the Birmingham tube-making firm of Grice, Grice & Sons, which was owned by Williams, Foster & Co. The public rhetoric of the new firm was upbeat. It was anticipated that "by the fusion of the trades in which we are jointly interested the new company will be in a far stronger position in those trades than either company individually has been in the past," and that by pooling its technical and administrative staff, "men of outstanding merit will henceforth be working together in the closest co-operation with a common aim, instead of as heretofore in active rivalry."[115] In reality, the merger was driven by an urgent need to cut costs and streamline the businesses. At a special meeting of the directors of Vivian & Sons in May 1924, where the practicalities of the new organization were discussed, it was acknowledged that, while in the short term Williams, Foster & Co. would continue to run Morfa and Middle Bank and Vivians their works at Hafod and Taibach, "ultimately it was proposed to close down Middle Bank and Margam [Taibach] with a view to general concentration."[116] The arrangement also dealt the final blow to copper smelting in the Lower Swansea Valley with the decision to cease this part of the Vivians' business, "having regard to the comparatively small outputs, lockup of capital and general difficulties as regards ore supplies."[117]

This decision to end copper smelting in the Lower Swansea Valley was not quite the end of Swansea Copper. As the old family firms disappeared and the number of works active in the trade shrank back to a core of premises around Landore, it was less an abrupt end than a fading from view of some of the most familiar landmarks in Swansea Copper's two hundred-year history. In its new configuration, "Hafod-Morfa" continued its metal fabrication operations in its complex of mills and workshops on the west bank of the River Tawe, the site that had been at the historic heart of the world copper industry for so long. Manufacturing of copper components there continued well into the post-1945 period for customers in the locomotive industry as well as firms of boiler makers, manufacturers of electric cable, engineering businesses, brewers, and distillers.[118] As it did so, its connection to the Swansea Copper of old grew ever weaker. Its identity changed as it passed through a succession of different commercial hands. British Copper Manufacturers became part of the metals division of Imperial Chemicals Industries (I.C.I.) in 1928, before being subsumed by Yorkshire Imperial Metals after the Second World War.[119] It was a sign of just how far the industry had shifted away from its origins in the reverberatory furnace technology of the early eighteenth century. While the establishment of smelting in North America, Chile, and South Australia in the middle decades of the nineteenth century had owed something to the Swansea example, the same could not be said for some of the key developments in copper production later on. The growth of the chemical industry in Britain and Europe instigated a much sharper departure from the traditional Welsh Process, with the development of new methods of extracting copper from pyrites. Demand in the electrical industry was a further shift, to high-conductivity metal from electrolytic refineries. If Swansea's ascent to the top of the global copper hierarchy had been founded on its geographical benefits as a location for smelting rich sulfide ores with coal, its loss of status was just as much a story of the undermining of these locational advantages and the rise of other centers more suited to new processing methods.

The Lower Swansea Valley after Copper

For Swansea, the town that had been so profoundly shaped by copper for two centuries, the shrinking back of the industry in the twentieth century was traumatic and painful. Much of the original corridor of copper along the banks of the River Tawe was abandoned and derelict by the 1960s (figure 6.5). The remaining Hafod-Morfa operations ceased in 1980. As the smelting works and processing plants closed their doors, so the cumulative effect of

Figure 6.5. White Rock copper works in the 1950s.
City and County of Swansea: Swansea Museum Collection, SM1985.206

two hundred years of intense metalliferous activity on the soil, air, vegetation, and landscape was exposed to public scrutiny. By the 1960s, chroniclers of Britain's postwar industrial decline had settled on the town as the archetypal symbol of damage and decay. "Nowhere in Derelict Britain," concluded one writer, "is there a more dismaying example of man creating wealth while impoverishing his environment than in the Lower Swansea Valley. A grey-black inverted triangle of 1,200 acres . . . the most concentrated and uninterrupted area of industrial dereliction in Britain."[120] The poisonous acreage and abandoned buildings of the former "Copperopolis" were not just an eyesore, and a visual reminder of economic decline; they also presented a practical obstacle to regeneration. Derelict land was a growing problem, with over forty thousand hectares designated derelict in England and Wales in 1964, rising to over fifty-eight thousand hectares ten years later.[121] The need to reclaim these areas and render them fit for use was a political priority in 1960s Britain, where demand for land for house building and road construction was being driven by changing residential, recreational, and demographic patterns. In Swansea, it was an alliance of local interest groups that mobilized to tackle the problem.

In 1961 a project team was established, made up of Swansea council members, university academics, and representatives of local groups, to survey and report on the condition of the former copper corridor and its prospects for redevelopment. Together they inaugurated an ambitious cleanup and landscaping plan, commencing in the early 1960s with an extensive tree-planting program involving local schoolchildren and community groups. The project team's survey work culminated in a report, published in 1967, that envisioned the zoning of the entire district into areas for new housing, recreation, new industries, and commercial development, all connected with the construction of a new valley access road.[122] It gave hope that, with careful planning and the creation of investment opportunities, new life could be breathed into the area. Yet its vision of the future had no place for remembering the industrial legacy of copper. The chimney stacks and furnace halls of the former copper works were an integral part of the blighted, toxic landscape; their removal was part of the process of reclamation and renewal.

This view of derelict industrial infrastructure was far from unique to 1960s Swansea. In the vast Butte-Anaconda mining and smelting district of Montana, also facing precipitous postwar decline, residents felt similarly burdened by the "huge toxic liability" of mining waste and contamination left behind as its mines and smelters wound down their operations, though there were earlier efforts there to document the former industrial importance of the district, notably with the opening of a museum on the site of a former mine in 1965.[123] In Swansea there was little pause for reflection. Site clearance was the overarching priority and, to this end, a local division of the Territorial Army, stationed in the town, was invited to use the ruined copper works sites for explosives and demolition training. The idea that former industrial buildings had any intrinsic heritage value had little traction in Britain in the 1960s. Swansea's former copper works were branded "hideous buildings," with the one notable exception of "the oddly elegant arches of the eighteenth-century White Rock Copper, Lead and Silver Works, which many people feel should have been preserved as a monument," though even this was not spared from the demolition team.[124] Subsequent events served only to harden the resolve behind such clearance work. The disaster in the Welsh village of Aberfan in October 1966, when a deadly torrent of coal waste engulfed a primary school and adjacent houses with horrifying loss of life, brought a new, grim determination to the sense that clearance and cleanup of derelict industrial sites was a public priority. Small wonder that the Lower Swansea Valley project became a model example of how to make safe and regenerate areas blighted by industrial decline.[125]

By the turn of the millennium, a small cluster of buildings on the western side of the River Tawe, consisting of a single chimney stack and engine shed of the former Vivian & Sons Hafod works, and buildings once housing a laboratory and canteen on the Morfa works site, were virtually all that remained of the once-mighty industrial heartland of Swansea Copper. With the surrounding area newly green and decontaminated, and home to a new football stadium, retail outlets, and a park-and-ride facility serving the city center, what little remained of Copperopolis took on a new significance. No longer part of an embarrassing black spot, darkening the reputation and future prospects of the town, the surviving structures on the Hafod-Morfa site acquired new meaning as a fragile remnant of Swansea's important place in the global story of copper. Meanwhile, the example of successful heritage regeneration projects at former coal mines and ironworks elsewhere in South Wales helped to demonstrate the development potential of historic industrial infrastructure.[126] A 1992 report into industrial archaeology in Wales demonstrated how far attitudes had shifted in a few short decades. It made the case for the "architectural dignity" of old industrial buildings and argued that "re-use or adaption for housing, offices or new light industries can nowadays make greater economic sense than . . . demolition."[127]

In 2012 the City and County of Swansea looked afresh at its surviving copper quarter, albeit a tiny fragment of what it had once been. It issued a marketing brief aimed at attracting new commercial investment to the former Hafod-Morfa site, which it defined as a twelve-and-a-half-acre riverside location containing twelve listed historic structures.[128] The restoration of these buildings—an unthinkable prospect in the 1960s—was now back on the agenda, and their potential to attract interest from businesses traditionally linked with copper or involved in new copper-related technologies was viewed as key to the future potential of the site. With the combined attention of the city council, the university, local businesses, and active volunteer groups once again focused on the Lower Swansea Valley, but this time with a shared interest in the possibility of using industrial heritage as a lever for regeneration, the prospect for Swansea's remaining copper assets seems bright. A century after the end of copper smelting in Swansea, the industry's time to reclaim a foothold in the valley it once dominated finally may have come.

Swansea Copper in World-Historical Perspective

Copper is humankind's oldest metal. Evidence gathered at sites in Iran and Serbia dates the earliest smelting to seven thousand years ago.[1] It is also one of the most widely worked metals. The smelting of copper ores has been practiced in Eurasia, sub-Saharan Africa, and the Americas since prehistory.[2] Swansea Copper therefore needs to be thought about against a backdrop of big time and big space. Doing so will allow us to highlight the Swansea District's singularity; it will also help puncture some of the more hyperbolic assertions made on behalf of the Welsh copper sector, of which there have been plenty. It is often claimed, for instance, that 90 percent of the world's copper was smelted in the Swansea Valley in the nineteenth century.[3] It was not; the percentage was far lower.

Swansea Copper as Revolution?

Swansea Copper *did* represent a radical break; of that there is no doubt—the use of coal technology was genuinely epochal. Yet Swansea Copper was not the first great upsurge in copper production in human history. Earlier phases of major expansion are recorded in the "frozen archive" of the Greenland ice core. The atmospheric fallout from early metallurgy in the northern hemisphere, captured in sedimented ice, indicates definite spikes in copper production ahead of the age of Swansea Copper. There was one at the very beginning of the Common Era and another in the eleventh century CE.[4] The first of these, environmental scientists have conjectured, corresponds to the spread of copper coinage through the ancient Mediterranean world under the aegis of Rome. The second uptick in atmospheric pollution has been tentatively ascribed to another imperial power that issued a copper currency: Song Dynasty China. Imperial collapse (in the case of Rome) or the vagaries of monetary policy (in the case of China) meant that neither of these early bursts of copper production was sustained.[5]

A third phase of expansion began in Europe at the tail end of the Middle Ages. Output did not rocket in the way that it had under the Roman Republic, but nor did it tail off. Despite regional ups and downs, European copper production was on an upward trajectory from the 1400s onward. This fact needs to be remembered. Swansea Copper did not irrupt into a torpid industrial landscape; it became part of a complex production network that had known three centuries of aggregate growth. Copper from early modern Mansfeld supplied brass producers in Aachen and Stolberg; Tyrolean copper descended from Alpine heights to upper Germany and northern Italy; Hungarian copper was routed to Mediterranean markets via Venice, and to Iberia via Danzig and Hamburg. Admittedly, the development of mining and smelting in central and eastern Europe was an uneven process, sometimes halted by crises of overproduction (as happened in the 1510s), sometimes held back by falling silver prices (something to which the miners of argentiferous ores in Mansfeld and Hungary were vulnerable), and sometimes arrested by the suppression of civilian demand during the endemic wars of the period.[6] Yet shortfalls in some areas were counterbalanced by surges in others. Thus, the decline of the central European mining zones in the seventeenth century was offset by the boom in Swedish mining, centered on the great mine at Falun, and the growth of copper production in Norway.[7]

The greater availability of copper in the early modern world was linked to the growth of seaborne trade and the closer integration of global markets. The experience of sub-Saharan Africa illuminates some of the processes at work. West Africa is poorly endowed with cuprous ores, so poorly that copper enjoyed the status of a precious metal. Demand was accordingly high among elite groups. Historically, this demand was met by a mix of local resources and material brought in via the trans-Saharan caravan trade. It was a trade that was underway in antiquity, making West Africa part of a far more extensive commercial network that connected the ancient Mediterranean world to central Asia and the Indian Ocean.[8] Trans-Saharan shipments increased in the Islamic era, drawing on ore deposits in North Africa and farther afield. Brass was particularly sought after because there were no known or workable zinc deposits in sub-Saharan Africa.[9] Over two thousand brass rods were found with the lost caravan of Maden Ijâfen, carbon dated to the twelfth or thirteenth century, when it was uncovered in the Mauritanian desert in the 1960s.[10] There was a limit to what could be done by trans-

Saharan means, however, which is why the arrival of Portuguese traders on the West African coast in the 1400s was transformative. They found a ready market for European copper and brass, one that they served with material sourced in central Europe and funneled through Antwerp. Portuguese agents in Flanders contracted for seventy-one thousand manillas in 1494–1495 alone.[11] This was to anticipate by two centuries the brasswares and Guinea rods that Swansea's copper works would launch into the Atlantic slave vortex. Producer goods, it is sometimes forgotten, have been traded over immense distances for centuries. The role of Swansea was not to initiate the global movement of copper but to amplify it massively.

When the Portuguese made the first tentative European entry into the flourishing world of South and Southeast Asian maritime commerce, copper went with them. Striking confirmation of this is furnished by the Oranjemund shipwreck, the remains of a Portuguese *nau* that foundered on the Namibian coast in the 1530s. Nearly eighteen tons of copper has been recovered from the site, in ingots bearing the seal of the Fugger family of Augsburg.[12] Portugal's eagerness to access Asian markets was entirely understandable; the Indian Ocean world boasted a product range of unmatched size and sophistication. Since copper was the stuff of numerous Asian currencies, there was every incentive to carry it eastward as a means of paying for Asian goods. European copper therefore joined an already busy circulation of the red metal in its unwrought and manufactured forms. That circulation quickened in the seventeenth century as mines in Rajasthan and central India began to be worked out.[13] Japanese copper, famed for its purity and peerless luster, filled the gap, especially after the Dutch established themselves as intermediaries carrying copper from Nagasaki to Indian Ocean ports. The take-up of Swansea Copper by the East India Company in the mid-eighteenth century must be seen in this context. It was a notable development for Swansea's copper companies. It was less momentous in South Asia; the arrival of Indiamen carrying Welsh copper merely added to the churn of Asian metal markets. Moreover, Welsh copper did not command respect. It was seen as "no more than a raw material for brass and artillery"; the "more greatly prized Japanese copper fetched higher prices in Bengal."[14]

For much of the early modern period, to reiterate points made in chapter 1, the British Isles were almost entirely dependent on copper from long-established mining zones elsewhere. The adoption of the reverberatory furnace in the 1680s and 1690s signaled the emergence of an entirely new industrial sector. In that sense, Swansea Copper lends support to those economic historians who detect a shift of labor out of agriculture and into industry well ahead of the Industrial Revolution as traditionally conceived.[15] The copper industry's contribution to this process should not be overestimated, however. The early works in the Bristol region employed a few dozen workers at most. Indeed, the copper industry in Britain, even at its Victorian zenith, employed relatively small numbers, sinking "into insignificance by the side of the iron trade."[16] Swansea Copper was capital intensive, not labor intensive, and although the masters were always keen to supress wage levels where they could, the payments made to furnacemen were a relatively minor component of the industry's costs. Swansea Copper provides, therefore, little succor to those historians who see a high-wage regime in Britain as a stimulus to labor-saving technological change, and thus the secret of Britain's industrial precocity.[17] In recent years a good deal of scholarly energy has been devoted to expounding or refuting the high-wage thesis. The debate is, however, dominated by textiles, particularly the question of why cotton spinning occasioned so much inventive activity.[18] Swansea Copper offers a different perspective. It reveals the transformational effects of mineral energy. Investors were attracted to the Swansea District by the abundance of coal; they were not driven there by the high wages paid out elsewhere.

Coal was fundamental, yet the new coal-fired industry was slow to cohere. Fifty years after the establishment of copper smelting by reverberatory in the Avon and Wye Valleys, the British continued to import substantial volumes of brass and unwrought copper. In the 1730s, brass battery wares from Holland still featured prominently on the British market. As for copper, perhaps a third of British output in the mid-1730s was accounted for by imported North African copper (Barbary copper) that was refined at British works.[19] Be that as it may, the steepening trajectory of Swansea Copper in the 1750s and 1760s was truly remarkable. British production of smelted copper in the 1740s hovered around the one thousand-ton mark. By 1775, it had quadrupled. By the end of the American Revolution, the four thousand tons produced in 1775 had jumped to seven thousand tons. The Industrial Revolution has no more dramatic tale to tell. And yet Swansea was not the only growth point. Russia followed an alternative developmental path (as noted in chapter 1), one that remained true to vegetable fuel, making use of the enormous forest resources of the Urals. The Chinese Empire followed suit. Copper was needed for the Qing Dynasty's mints, so the authorities made huge efforts to develop mining and smelting on the Empire's turbulent southwest frontier. The results were stunning. Yunnan province was producing nine thousand tons of smelted copper at its zenith in the 1760s. Neighboring Sichuan contributed a further one thousand tons annually, on average, in the second half of the eighteenth century.[20] Here was an achievement that matched, indeed overmatched, events in Wales. From an Asian perspective, the rise of the Swansea District does not appear quite as transcendent as Eurocentric historians have long assumed.

Even in Europe, triumphalist claims on Swansea's behalf need to be tempered. Production in the Swansea and Neath Valleys stormed upward in the eighteenth century. That much is incontrovertible. Yet the old heartlands of European copper making mostly endured. Although Falun never recovered from the calamitous cave-ins of the late seventeenth century, new scholarship on copper making in the Habsburg monarchy detects fuel-saving initiatives and product innovation in the eighteenth century, perhaps even a growth in output.[21] Central Europe's mining districts were no longer dominant but they were not necessarily decadent. The longevity of Mansfeld, for example, was extraordinary, so much so that it was hailed "as an exception to the ordinary laws of copper-mining."[22] Like all of the older mining districts, Mansfeld had exploited local forest resources. In the nineteenth century, with timber reserves under strain, Mansfeld smelters switched to coal. To do so was an acknowledgment of the historic breakthrough made by Swansea Copper. Mansfeld was not Swansea in facsimile, however. Lacking local coal, Mansfeld's smelters had to bring in their fuel by rail from the Ruhr.[23] The processing techniques used were, as a consequence, as sparing of coal as they could be. Mansfeld was therefore as well adapted to its particular resource endowment as the Swansea District was to its own, as one nineteenth-century commentator noted: "[The] Mansfeld process is characterized by the expenditure of a large amount of time and labour, with economy of fuel, the English [*sic*] process is distinguished by great economy of time and labour, with a comparatively large expenditure of fuel. It does not hence follow that in either case the one process is to be considered superior to the other, because each is best suited to the locality and circumstances under which it is conducted."[24]

Swansea Copper as Capitalism

To focus on technology, though, is to overlook a less remarked upon but nonetheless world-tilting aspect of Swansea Copper: its *capitalist* character. Earlier forms of copper production were embedded within tributary networks (in Mesoamerica), or rooted in seigneurial privilege (across most of early modern Europe), or reliant on imperial authority (as was the case in the Russian and Chinese empires). Swansea Copper, by contrast, arose in the context of generalized commodity production in which self-reproduction, each time on an extended scale, was the copper companies' sole purpose. Profit was paramount, and the plowing back of profits into expanded production a distinguishing feature. Elsewhere, smelters were built to serve the needs of imperial mints, or to harmonize with the cameralist priorities of early modern states. They served needs that were specific and/or limited. Swansea's reverberatories served needs that were nonspecific and that knew no limit.

That said, Swansea Copper operated within a very particular political economy, the "actually existing capitalism" of eighteenth-century Britain.[25] Swansea Copper was part of the new institutional landscape that took shape after the Revolution of 1688. Innovative mechanisms for extending credit and sharing risk enabled City bankers and merchants to mobilize unprecedentedly large sums on the emergent London money markets. Some of this capital was channeled into southwest Wales, much of it packaged in joint-stock form. The pioneer of copper smelting in the Neath Valley, Sir Humphry Mackworth, twice resorted to joint-stock vehicles, once with the Company of Mine Adventurers of England (1698) and then the Mineral Manufacturers of Neath (1713). The first smelting venture in the Swansea Valley, the works at Llangyfelach, was a more traditional partnership, but those involved were deeply rooted in the new world of metropolitan high finance. One of them, Edward Gibbon, was a prominent promoter of the South Sea Company (and barely survived its wreck in 1720); another, the international merchant Richard Lockwood, was deputy governor of the Royal Exchange Assurance Company.

Such men had the means of amassing money, which gave them in turn the means of commandeering mineral energy. The reverberatory furnace, with its insatiable appetite for coal, was able to flourish in South Wales because of an institutional quirk that made coal readily available. In England and Wales (unlike much of continental Europe) it was an accepted legal principle that mineral deposits, gold and silver excepted, were the property of whoever owned the land beneath which they lay. And because the distribution of land in England and Wales was skewed so heavily toward large landowners, those with a title to subterranean resources were also those who were financially best placed to extract them. The gentry of southwest Wales were not well-off by English standards but they were wealthy enough to build up a significant coal trade (and powerful enough to extinguish the rival claims of smallholders and squatters).[26] Onto this already flourishing coal trade the moneybags promoters of the new copper industry of the early eighteenth century were to batten.

Capital and coal, then, were on hand. So too were new markets, courtesy of the militarism and colonial aggression that was so marked a feature of eighteenth-century British political economy.[27] To revisit a point made in chapter 1, the initial success of Swansea Copper was intimately linked to Atlantic slavery. While the Caribbean sugar sector was the *key* external market in the first half of the eighteenth century, copper played an important role in the acquisition of enslaved Africans too. It was fitting that one of the most dynamic movers in the new copper industry of the 1690s, William Dockwra, was a slave trader; fitting too that his works at Upper Redbrook featured a mill for making "negroes," the copper rods that acted as currency on the Guinea coast.[28] Members of the Coster family, associates of Dockwra at Redbrook and subsequently active in their own right in the Swansea District, also invested in slave trafficking. They sponsored several triangular voyages linking Bristol, West Africa, and South Carolina in the 1730s.[29] Swansea Copper was an imperial project from the outset and long remained one.

Another token of Swansea Copper's capitalist character was its use of commodified labor. Workers in the Swansea District were paid wages, usually in cash. Although Swansea Copper had no qualms about cohabiting with chattel slavery in distant mining districts like El Cobre, the core relationship between the smelting companies and those who labored for them was contractual, albeit highly asymmetric, given that contracts were framed within a "masters and servants" legal tradition that greatly advantaged the masters. That much is plain, but much remains to be done in exploring labor relations within the Welsh copper industry. We have made a beginning in chapters 4 and 6, dwelling upon the crisis years of the 1840s and the shifting composition of the workforce toward the end of the nineteenth century. The questions that have been resolved are, however, far outweighed by those that await resolution. For one, we still know very little about where new workers were found during Swansea Copper's periods of headlong expansion. And while the strike of 1843 shines a light on workplace tensions in the early Victorian age, for two to three decades thereafter the picture is shadowy and indistinct. Likewise, we must guess at changing gender relations in the mid-nineteenth century, or reactions to the new technologies that were taken up between the 1860s and 1880s.

The Limits to Swansea Copper

Swansea Copper was revolutionary, no doubt, but great revolutions, having been accomplished, need no repetition. The Welsh Process was hybridized repeatedly; it was seldom cloned. Perhaps only the works established in Baltimore in the 1840s can be considered as faithful imitations of the Swansea model. Other early efforts at transplanting coal-fired copper smelting into new contexts—those in South Australia, for example, or the works of the Mexican & South American Company in Chile—were conscious attempts to condense the Welsh Process, not to replicate it. Indeed, it was not long before the Welsh Process began to be sidelined. In the 1870s and 1880s, as the American West became the new frontier of global copper production, the accent was on pneumatic and electrical technologies that flourished best when conducted on an epic scale ("a general scale of operations too large to be in harmony with any but an exceptional smelting plant").[30]

Swansea Copper was ushered toward historical redundancy by the extraordinary transformation in mining methods that began in the American West at the start of the twentieth century. The Cornish miners who served Swansea Copper in its heyday engaged in selective extraction: they identified a metalliferous vein and pursued it underground until it tapered off. It had been the way of miners since prehistory. The nonselective mining methods that made their debut in the United States on the eve of the First World War were radically different. Rather than laboriously detaching richly mineralized materials from commercially worthless country rock, those who pioneered nonselective techniques sought out geological features in which low-grade ore—ore that was barely distinguishable from the country rock—was present. Rather than sink shafts or drive adits into mountainsides, the new-model mining operators chose to remove the lightly mineralized rock wholesale. By using what have been called "mass destruction" methods (a mining analogue of Fordist mass production), ore bodies with a copper content as low as 2 percent could be made to pay. High-speed, high-volume throughput was the key to success. Vast excavated bowls of the sort that began to pock Utah were the result: industrial amphitheaters shaken by dynamite and the rumble of giant rock-shifting vehicles.[31] Entire mountains were disassembled in the process, the rock pulverized, and copper extracted from the debris by means of gravity concentrators and flotation cells.

These new forms of mining, just like the newer forms of smelting that now prevailed in the United States, were Pharaonic in scope and ambition. They demanded capital investment on a staggering scale and encouraged vertical integration, bringing together ore extraction, the command of water and power sources, and smelting within a single corporate envelope. Swansea Copper may have been capitalism's first copper, but capitalism, like rust, never sleeps. Organizationally as well as technologically, the copper industry of the early twentieth century had left Swansea far behind.

Swansea Copper began the industrialization of copper production. Fossil fuel allowed for the smelting of copper on a historically unprecedented scale. The breakthrough was fundamental: the umbilical link between smelting and local ores was severed. In its place came a reliance on seaborne ores. Anywhere with ore lodes tolerably close to a harbor could now be harnessed to Swansea Copper. Extendable supply chains were a necessary component of the Swansea model, because the use of coal meant that smelting capacity could run ahead of ore supplies—so far ahead, in fact, that Swansea copper masters scoured the earth for additional supplies in the 1830s and 1840s. The gulf between mine and smelter yawned ever wider. Swansea Copper became a transoceanic phenomenon.

Despite the huge geographical distances involved, this was industrialization of a characteristically British sort. Swansea Copper exploited mineral energy and human muscle power. It was a potent mix. Furnacemen in the Swansea District and miners in Cornwall are the best-remembered human participants in this process. They are revered for their hard-won expertise: furnacemen for managing the elemental fury encased within reverberatories, miners for discerning the trend of a mineral vein in the stygian depths. Less celebrated are the women and juvenile workers whose labor, whether devoted to the relentless spalling of ore or the barrowing away of slag, was no less sapping.

The copper industry that developed in the United States in the last decades of the nineteenth century drew upon British precedent but quickly transitioned into something distinctively new. The new technological paradigm that emerged in the American West collapsed the distance between mine and smelter, reinstalling the age-old model spoken of by Frédéric Le Play. It also brought about a new labor regime: managers with technical qualifications took the place of men who had been schooled in the vernacular practices of Swansea Copper, while furnacemen who acted on their materials with an artisanal directness gave way to workmen who used pneumatic and electrical powers. The American West was now the arbiter of global best practice. Veterans of the Welsh scene—like Thomas Nicholls of the

Cape Copper Company—conceded the point by introducing American-inspired management methods. Elsewhere, the influence of Wales receded fast. The superintendent of an Australian smelting works in 1870 might still be a Welsh patriarch who had, as a young man back in the '30s or '40s, tended a furnace in the Swansea District, but by the start of the twentieth century his successor was "far more likely to be a young American with a German surname and an engineering diploma from Arizona or Michigan."[32]

Swansea began the industrialization of copper, but it could not complete it.

The Welsh Process

Frédéric Le Play made an exhaustive analysis of the Welsh Process in the 1840s, when Swansea Copper was at its zenith. In his *Description des procédés métallurgiques employés dans le Pays de Galles pour la fabrication du cuivre* he denominated ten different operations. He made a summary listing of these first in French, "then by the names used in the locality" (100). The list is reproduced below.

Additional notes are taken from Sheridan Muspratt's *Chemistry: Theoretical, Practical, and Analytical, as Applied and Relating to the Arts and Manufactures* (Glasgow: William Mackenzie, 1860), which followed Le Play closely, identifying the great virtue of the Welsh Process as its ability to extract copper from sulfur-rich ores. It had to be done patiently, step-by-step:

> By the action of the heat of a reverberatory furnace, considerable quantities of sulphur are expelled in the first operation, leaving the copper combined with a minimum of that element, which it tenaciously retains. In the next, the whole or chief part of the iron is removed by fusing the ores with some silicious matters. . . . In this fusion the copper manifests a greater affinity for the sulphur than the iron, which is readily oxidised by the air, and the oxide thus produced is taken up by the silicious matters, leaving the matt of sulphide of copper behind, rich in metal, but still containing some sulphide of iron. By a further roasting with other ores, such as the oxide or carbonate of the metal, containing a sufficient amount of silica or quartz, a slag is thrown off, in which is all or most of the iron, leaving the matt of sulphide of copper richer than what was produced in the foregoing processes. It is thus by successive calcinations that all the iron and sulphur are removed, leaving a fusible mixture called black copper, which is afterwards refined. . . .

Le Play's ten stages were:

I. *Grillage des minerais sulfurés (pauvres et de richesse moyenne)*: Calcination of the ores

   MUSPRATT: "The calcination of the ores by which the metals are reduced to a minimum degree of sulphuration."

II. *Fabrication de la matte bronze, ou fonte des minerais pauvres (bruts et grillés)*: Melting for coarse metal
MUSPRATT: "Melting the above product [calcined ore] for coarse metal; in this operation, a quantity of fresh mineral, rich in sulphide of copper, is added to the calcined product of the first roasting."

III. *Grillage de la matte bronze*: Calcination of coarse metal
MUSPRATT: "A similar purpose is here entertained as in roasting the crude ore [stage I], namely the expulsion of the sulphur, and the production of oxidised bodies."

IV. *Fabrication de la* matte blanche ordinaire *ou fonte de la matte bronze grillée avec les minerais riches*: Melting for white metal
MUSPRATT: "Melting of the calcined crude metal with rich ores of the fourth class [copper oxides and carbonates], to produce *white metal*. . . . The procedure in this stage of the work is analogous to the fusion of the *matt* in the second process. It has for its object the removal of the iron, and the production of pure sulphide of copper. . . ."

V. *Fabrication de la* matte bleue, *ou fonte de la matte bronze grillée, avec les minerais grillés de richesse moyenne*: Melting for blue metal
MUSPRATT: "Fusion of the white metal with roasted minerals, rich in copper, by which *blue metal* is produced."

VI. *Fabrication des* mattes blanche et rouges (de scories)*, ou fonte de scories riches des operations IV, VII et VIII*: Remelting of slags
MUSPRATT: "In this stage, the slags resulting from the preceding and the two succeeding fusions, and which are rich in oxide of copper, as well as some rich sulphide of copper from certain ores, which, however, are free from injurious substances, are treated."

VII. *Fabrication de la* matte blanche-extra, *ou rótissage de la matte bleue*: Roasting of white metal
MUSPRATT: "*Blue metal* is here converted into white metal, by the agency of the air; and the chief or entire part of the remaining iron is removed, by forming a fusible silicate toward the close of the calcination."

VIII. *Fabrication des* matte - régules *ou rótissage de la matte blanche-extra*: Roasting for regulus
MUSPRATT: "This seems to be only a repetition of the preceding treatment, and is conducted in almost the same manner. It constitutes the last of the series called the *extra process*, and yields a substance which like that resulting from *operation four* in the ordinary mode, is ready for the calcination by which the metal is obtained."

IX. *Fabrication du cuivre brut, ou rótissage de la matte blanche ordinaire, des mattes-regules et des fonds cuivreux*: Roasting
MUSPRATT: "Preparation of crude copper from the ordinary white metal regulus, *et cetera*." It was here that the white metal produced in operation IV was combined with materials yielded by the *extra process*. The outcome was *blistered copper*.

X. *Raffinage du cuivre brut et préparation du* cuivre malléable: Refining and toughening
MUSPRATT: "Refining, and production of tough malleable metal . . . so as to bring it to that state adapted for the mechanist."

GLOSSARY

The following are the main works consulted in the definition of terms included in the glossary: John Morton, "The Rise of the Modern Copper and Brass Industry in Britain, 1690–1750" (PhD thesis, University of Birmingham, 1985); H. C. H. Carpenter, "Progress in the Metallurgy of Copper," Lectures 1 and 2, *Journal of the Royal Society of Arts* 66, no. 3398 (January 1918); Andrew Ure, *A Dictionary of Arts, Manufactures and Mines; Containing a Clear Exposition of Their Principles and Practice* (New York: D. Appleton and Co., 1842); J. H. Vivian, "An Account of the Process of Smelting Copper as Conducted at the Hafod Copper Works near Swansea," *Annals of Philosophy* new series 5 (1823); R. O. Roberts, "The Smelting of Non-ferrous Metals," in *Glamorgan County History*, volume V: *Industrial Glamorgan from 1700 to 1970*, ed. Glanmor Williams and A. H. John (Cardiff: Glamorgan County History Trust Ltd., 1980); R. R. Toomey, "Vivian and Sons, 1809–1924: A Study of the Firm in the Copper and Related Industries" (PhD thesis, University of Wales, 1979); E. D. Peters, *The Practice of Copper Smelting* (New York: McGraw-Hill Book Co., 1911); W. Halcrow, "The Carriage of Heavy Ore Cargoes," *Journal of the Royal Society of Arts* 83 (April 1935); J. C. Symons, "The Mining and Smelting of Copper in England and Wales, 1760–1820" (MA thesis, Coventry University, 2003); A. H. John, "Iron and Coal on a Glamorgan Estate, 1700–1740," *Economic History Review* 13, no. 1/2 (1943): 96.

Barbary copper. Coarse, unrefined cake copper from the "Barbary Coast" of North Africa, this was imported by Swansea copper smelters in the early eighteenth century to supplement domestic supplies of copper ore from Cornwall.

Battery ware. Manufactured products made from sheets of copper and brass and shaped using water-powered hammers. Battery ware was imported from Holland in the seventeenth and early eighteenth centuries, but Bristol and London became important British centers of production with mills located near the rivers Avon and Thames respectively.

Bessemer converter. Apparatus invented in the 1850s for the production of steel from molten iron. The Bessemer process involved the blowing of air through the molten metal and was adapted for use in copper smelting on a commercial scale by Pierre Manhès in Eguilles, France, in the early 1880s.

Best Selected copper. A term used in the marketplace (often abbreviated to B.S.) to denote copper of the highest purity, usually produced in the form of ingots and used in the manufacture of brass and other alloys. The term was originally used to refer to the purest form of copper obtained by the Welsh Process, but it continued to be used after the development of other processing methods to refer to refined copper of up to 99.7 percent metal content.

Blister copper. A form of copper obtained during the ninth stage of the Welsh Process of copper smelting, before refining. Blister copper was so called because of the formation of blisters on the surface of the metal caused by the escape of sulfur dioxide. By the 1880s, this stage of the Welsh Process had been superseded, and blister copper could be produced in greater bulk using a Bessemer converter.

Calamine. A carbonate of zinc used in conjunction with copper in the production of brass.

Calcining. The process of reducing a mineral through burning or roasting. This term was not particular to the practice of copper smelting but was used to describe various phases of the Welsh Process, as well as the types of furnaces used in the operation.

Clinker. Furnace slag.

Copper bark. Three-masted sailing vessels commonly employed to ship copper ore to Swansea. These were adapted for the safe carriage of heavy ore cargoes with modifications to their holds so as to raise the center of gravity and increase the stability of the vessel, even in rough seas.

Copper bottoms. Products including flat bottoms and raised bottoms were produced by Swansea copper firms for use in the manufacture of products such as stills, pans, and other equipment used by food and drink producers. (The term is not to be confused with furnace bottoms: residues with usable metal content accumulating in the bottoms of furnaces after repeated smeltings; or copper sheathed ships—see below—which were sometimes referred to as being copper-bottomed.)

Copper sheathing. Sheets of copper used to protect the hulls of ships from corrosion caused by the *teredo navalis* (shipworm). Developed following trials by the Royal Navy in the 1760s, it improved the sailing speed as well as the longevity of a vessel and was widely adopted by European fleets until the availability of cheaper alloys in the 1830s.

Copper shot. A granulated form of copper used in large quantities in brass making in a process pioneered by the Bristol Quaker, Nehemiah Champion, in the 1720s. Different forms of copper shot, including bean shot and feathered shot, were produced by Swansea smelting firms in the eighteenth and early nineteenth centuries.

Copper standard. The value of one ton of cake copper, the copper standard was used, along with other factors, including the yield of metal in a parcel of ore and the charges incurred in transporting and smelting ore, to determine the prices paid for copper ore at the ticketings.

Cupola furnace. A metal smelting furnace with a domed furnace roof. This term came to be used to describe the reverberatory furnaces in use in the copper industry in Britain from the late seventeenth century.

Direct refinery process. A modification of the Welsh method of copper smelting, patented by Thomas Nicholls and Christopher James at the Cape Copper Company, Briton Ferry, in the 1890s. It enabled smelters to bypass the roasting stage, thereby saving cost and time in the production process.

Electrolytic refining. The refining of metallic copper, in the form of anode plates, by dissolving it in sulfuric acid, then using an electric current to produce a precipitate of pure metallic copper (cathodes). The process yielded copper of very high purity and high electrical conductivity.

Guinea rods. Small lengths of copper, usually hammered into the required dimensions, and sold in the mid-eighteenth century to firms engaged in trade on the West African coast. They were often carried as part of mixed cargoes traded in exchange for slaves.

Japan copper. Six-inch lengths of copper with a reddish surface color caused by oxidization of the metal during the production process. They were produced for export to the East Indies.

Manillas. Small, horseshoe-shaped items produced in Swansea copper works from the 1720s. Usually cast in brass or a copper-lead alloy, they were popular as ornamental wear and currency. Like Guinea rods, these were sold to Bristol firms engaged in the West African slave trade.

Pyrites. A sulfide copper ore found in abundance in southern Spain. Originally mined for its sulfur content, its use in copper production was enabled by the invention of a "wet process" for the extraction of the metal content by the Glasgow chemist William Henderson in the 1860s.

Rabbling. The act of stirring, skimming, or raking material within a reverberatory furnace.

Regulus. A part-smelted form of copper, also known as copper matte, separated from the ore after an initial phase of furnace treatment. Regulus became an increasingly important form of copper imported into Britain from the mid-1860s onward, especially from Chile. Containing a higher metal content than ore, it was more cost effective to ship over long distances.

Reverberatory furnace. The defining technology of Swansea Copper, this furnace enabled the smelting of copper ore using coal rather than timber or charcoal fuel. It featured a partition separating the ore from the fuel during the smelting process, thus shielding the copper from exposure to impurities in the coal. Its other key features were a tall chimney and a sloping furnace roof that allowed intense levels of heat to reverberate over the ore during firing.

Swansea District. Part of the Bristol Channel coast of South Wales stretching from Llanelli in the west to Taibach in the east—a distance of some twenty miles—in which the major concentration of Britain's copper-smelting works were located for much of the eighteenth and nineteenth centuries. The district centered on the port of Swansea and the valley extending north of the town along the banks of the River Tawe.

Ticketings. Public sales of copper ore, these were held every two weeks in Cornwall, beginning in 1725, for the sale of Cornish ores. In the 1810s, ticketings commenced in Swansea for the sale of Welsh, Irish, and other overseas ores. The Swansea ticketings also functioned as a regular gathering point for the exchange of news and trading information between smelting firms.

Tribute system. A form of contract labor used in the Cornish mining industry whereby workers rented a "pitch" in the mine and were paid according to the selling price of the ore they raised. Tributers had to cover the cost of their tools and materials as well the cost of conveying waste to the surface of the mine.

Tutwork. Tutworkers, like tributers, were employed underground in Cornish metal mines, but were engaged in mine development work rather than ore raising. They were paid by the fathom for driving levels and sinking shafts, with their earnings subject to variability according to the nature of the ground being worked.

Welsh Process. Also sometimes referred to as the "English" process, this was the method of smelting copper in a reverberatory furnace, using coal as the fuel source. It involved subjecting the ore to repeated roastings in a ten-stage process that consumed some eighteen tons of coal per ton of copper ore. It was adopted in Swansea at the Llangyfelach works in 1717 and was expanded and perfected by subsequent firms in the district using locally mined coal.

Wey. The measure commonly used for quantifying coal in the western part of the South Wales coalfield. Although subject to local variation, it was considered equal to five tons in 1717.

Yellow metal. A copper-zinc alloy patented in 1832 by the Birmingham manufacturer George Frederick Muntz. Also known as "Muntz Metal," it was a cheaper alternative to copper sheathing and was taken up initially by merchant shipping before being adopted by the Admiralty.

NOTES

Abbreviations

BA	Birmingham Archives
BPP	British Parliamentary Papers
BUA	Bangor University Archives
CBS	Centre for Buckinghamshire Studies
Forbes Papers	Forbes of Callendar Papers, Falkirk Archives
GA	Glamorgan Archives
HMSO	Her Majesty's Stationery Office
NLW	National Library of Wales
RBA	Richard Burton Archives, Swansea University
RIC	Royal Institution of Cornwall
SCL	Swansea Central Library
TNA	The National Archives
WGAS	West Glamorgan Archive Service

Introduction

1. This estimate of Swansea's share of world output may seem conservative to some, as claims for a far higher share are often made. We have erred on the side of caution because the key statistical source, Christopher J. Schmitz's *World Non-ferrous Metal Production and Prices, 1700–1976* (London: Frank Cass, 1979), under-plays Chinese production, a major component of world output in the eighteenth century.

2. Frédéric Le Play, *Description des procédés métallurgiques employés dans le Pays de Galles pour la fabrication du cuivre* (Paris: Carilian-Goeury et Von Dalmont, 1848), 6–7.

3. A. Snowden Piggot, *The Chemistry and Metallurgy of Copper, including a Description of the Principal Copper Mines of the United States and Other Countries* (Philadelphia: Lindsay & Blakiston, 1858), 185.

4. Jonas M. Nordin, "Metals of Metabolism: The Construction of Industrial Space and the Commodification of Early Modern Sápmi," in *Historical Archaeologies of Capitalism*, ed. M. P. Leone and J. E. Knauf (Heidelberg: Springer, 2015), 249–272.

5. Clair C. Patterson, "Copper, Silver, and Gold Accessible to Early Metallurgists," *American Antiquity* 36, no. 3 (1971): 286–321.

6. Caroline Robion-Brunner, "L'Afrique des métaux," in *L'Afrique ancienne de l'Acacus au Zimbabwe: 20 000 avant notre ere—xviie siècle*, ed. Francois-Xavier Fauvell (Paris: Belin, 2018), 519–543.

7. Eugenia W. Herbert, *Red Gold of Africa: Copper in Precolonial History and Culture* (Madison: University of Wisconsin Press, 1984).

8. Kristof Glamann, "The Dutch East India Company's Trade in Japanese Copper, 1645–1736," *Scandinavian Economic History Review* 1 (1953): 41–79.

9. Chris Evans and Göran Rydén, *Baltic Iron in the Atlantic World in the Eighteenth Century* (Leiden: Brill, 2007); Chris Evans and Alun Withey, "An Enlightenment in Steel? Innovation in the Steel Trades of Eighteenth-Century Britain," *Technology & Culture* 53, no. 3 (2012): 533–560.

10. T. S. Ashton, *The Industrial Revolution 1760–1830* (Oxford: Oxford University Press, 1968).

11. C. A. Bayly, "'Archaic' and 'Modern' Globalization in the Eurasian and African Arena, c. 1750–1850," in *Globalization in World History*, ed. A. G. Hopkins (London: Pimlico, 2002), 47–73.

12. Giorgio Riello, *Cotton: The Fabric That Made the Modern World* (Cambridge: Cambridge University Press, 2013); Maxine Berg, *Luxury and Pleasure in Eighteenth-Century Britain* (Oxford: Oxford University Press, 2005).

13. J. J. Mason, "Arkwright, Sir Richard (1732–1792)," *Oxford Dictionary of National Biography* (Oxford: Oxford University Press, 2004), http://www.oxforddnb.com/view/article/645.

14. *Report of the Select Committee of the . . . Directors of the East India Company, Upon the Subject of the Cotton Manufacture of this Country* (London, 1793), 6, quoted in Giorgio Riello, "Cotton: The Making of a Modern Commodity," paper presented at ISECS 2015, Rotterdam.

15. Dafydd Tomas, *Michael Faraday in Wales: Including Faraday's Journal of His Tour through Wales in 1819* (Denbigh: Gwasg Gee, 1972), 36. See also John Scoffern, William Truran, William Clay, Robert Oxland, William Fairbairn, W. C. Aitkin, and William Vose Pickett, *The Useful Metals and Their Alloys, including Mining Ventilation, Mining Jurisprudence, and Metallic Chemistry Employed in the Conversion of Iron, Copper, Tin, Zinc, Antimony and Lead Ores; with Their Application to the Industrial Arts* (London: Houlston & Wright, 1866), 551.

16. Nuala Zahedieh, "Colonies, Copper, and the Market for Inventive Activity in England and Wales, 1680–1730," *Economic History Review* 66, no. 3 (2013): 805–825.

17. J. R. Harris, "Copper and Shipping in the Eighteenth Century," *Economic History Review* 19, no. 3 (1966): 550–568; Peter M. Solar and Klas Rönnbäck, "Copper Sheathing and the British Slave Trade," *Economic History Review* 68, no. 3 (2015): 806–829.

18. *R. R. Angerstein's Illustrated Travel Diary, 1753–1755: Industry in England and Wales from a Swedish Perspective*, ed. Torsten Berg and Peter Berg (London: Science Museum, 2001), 324; Joan Day, *Bristol Brass: A History of the Industry* (Newton Abbot: David & Charles, 1973), 199.

19. Kenneth Pomeranz, *The Great Divergence: China, Europe, and the Making of the Modern World Economy* (Princeton: Princeton University Press, 2000). See also Prasannan Parthasarathi, *Why Europe Grew Rich and Asia Did Not: Global Economic Divergence, 1600–1850* (Cambridge: Cambridge University Press, 2011).

20. E. A. Wrigley, *Continuity, Chance and Change: The Character of the Industrial Revolution in England* (Cambridge: Cambridge University Press, 1998). See also R. C. Allen, *The British Industrial Revolution in Global Perspective* (Cambridge: Cambridge University Press, 2009).

21. MS 3782/12/108/27, p. 78, Matthew Boulton papers, BA.

22. "Price Current of Hungarian Copper as Sold at the Imperial Warehouses in Trieste in 1781," MS 3782/12/90/79, BA.

23. For a discussion of Tokat, see Joseph Franel to Matthew Boulton, November 16, 1799, MS 3782/12/90/106, and memorandum on copper mines in "Netolia," MS 3782/12/90/107, BA.

24. David Killick and Thomas Fenn, "Archaeometallurgy: The Study of Preindustrial Mining and Metallurgy," *Annual Review of Anthropology* 41 (2012): 560.

25. This was a feature of the work of R. O. Roberts, the doyen of Swansea's industrial history. See his "Development and Decline of Copper and Other Nonferrous Metal Industries in South Wales," *Transactions of the Honourable Society of Cymmrodorion* (1956): 78–115, and "The Smelting of Non-ferrous Metals since 1750," in *Glamorgan County History*, vol. 5, *Industrial Glamorgan from 1700 to 1970*, ed. A. H. John and Glanmor Williams (Cardiff: Glamorgan County History Trust, 1980), 47–95.

26. The title of Ronald Rees's *King Copper: South Wales and the Copper Trade 1584–1895* (Cardiff: University of Wales Press, 2000) suggests a wide-ranging study. However, the book focuses more narrowly on the environmental damage wreaked by the copper industry in the nineteenth century.

27. John Morton, "The Rise of the Modern Copper and Brass Industry in Britain 1690–1750" (PhD thesis, University of Birmingham, 1985). Morton's supervisor, John R. Harris, made some penetrating observations on the eighteenth-century copper industry in his biography of the industrialist Thomas Williams (1737–1802), *The Copper King: A Biography of Thomas Williams of Llanidan* (Liverpool: Liverpool University Press, 1964). Regrettably, Harris never produced a book-length work of synthesis to succeed Hamilton's.

28. This was the fate of Morton's "Rise of the Modern Copper and Brass Industry." Some of the findings of Edmund Newell's "The British Copper Ore Trade in the Nineteenth Century, with Particular Reference to Cornwall and Swansea" (DPhil thesis, University of Oxford, 1988), were condensed into journal articles, but not all. His most important published statement is "'Copperopolis': The Rise and Fall of the Copper Industry in the Swansea District, 1826–1921," *Business History* 32, no. 3 (1990): 75–97.

29. For an exception, see H. V. Bowen, "Sinews of Trade and Empire: The Supply of Commodity Exports to the East India Company during the Late Eighteenth Century," *Economic History Review* 55, no. 3 (2002): 466–486.

30. For attempts to move on, see Chris Evans, "El Cobre: Cuban Ore and the Globalization of Swansea Copper, 1830–1870," *Welsh History Review* 27, no. 1 (2014): 112–131; Bill Jones, "Labour Migration and Cross-Cultural Encounters: Welsh Copper Workers in Chile in the Nineteenth Century," *Welsh History Review* 27, no. 1 (2014): 132–154; Louise Miskell, "From Copperopolis to Coquimbo: International Knowledge Networks in the Copper Industry of the 1820s," *Welsh History Review* 27, no. 1 (2014): 92–111.

31. A preliminary statement is Chris Evans and Olivia Saunders, "A World of Copper: Globalizing the Industrial Revolution, 1830–1870," *Journal of Global History* 10, no. 1 (2015): 3–26.

Chapter 1 • Copper in Baroque Europe

1. David Killick and Thomas Fenn, "Archaeometallurgy: The Study of Pre-industrial Mining and Metallurgy," *Annual Review of Anthropology* 41 (2012): 563.

2. Braziers were the makers of domestic copper wares. Coppersmiths made "Coppers, Boilers for the Brewers, and all manner of large vessels"; their work was "the largest and most laborious." R. Campbell, *The London Tradesman, Being a Compendious View of All the Trades, Professions, Arts, Both Liberal and Mechanic, Now Practised in the Cities of London and Westminster* (London: T. Gardner, 1747), 264.

3. Donald L. Fennimore, *Metalwork in Early America: Copper and Its Alloys from the Winterthur Collection* (Winterthur: Henry Francis du Pont Winterthur Museum, 1996), 20.

4. Carlo M. Cipolla, *Guns and Sails in the Early Phase of European Expansion 1400–1700* (London: Collins, 1965).

5. A convenient English-language overview of these developments is given in Ekkehard Westermann, "Copper Production, Trade and Use in Europe from the End of the Fifteenth Century to the End of the Eighteenth Century," in *Copper as Canvas: Two Centuries of Masterpiece Paintings on Copper 1575–1775*, ed. Michael K. Komanecky (Oxford: Oxford University Press for Phoenix Art Museum, 1999), 117–130.

6. Ian Blanchard, *Russia's "Age of Silver": Precious Metal Production and Economic Growth in the Eighteenth Century* (London: Routledge, 1989), chapters 1 and 2.

7. John C. Symons, "The Mining and Smelting of Copper in England and Wales 1760–1820" (MPhil thesis, Coventry University, 2003), 173, table 2.5.

8. Blanchard, *Russia's "Age of Silver,"* 189.

9. Henry Hamilton, *The English Brass and Copper Industries to 1800* (London: Frank Cass & Co. Ltd., 1967), 1–13.

10. Hamilton, *English Brass and Copper*, 13–20.

11. The dominant position of Swedish copper in Restoration England is a key theme of *Markets and Merchants of the Late Seventeenth Century: The Marescoe-David Letters, 1668–1680*, ed. Henry Roseveare (Oxford: Oxford University Press for the British Academy, 1991).

12. Hamilton, *English Brass and Copper*, 68.

13. Hamilton, *English Brass and Copper*, vii.

14. George Hammersley, "Technique or Economy? The Rise and Decline of the Early English Copper Industry, ca. 1550–1660," *Business History* 15, no. 1 (1973): 18.

15. John Morton, "The Rise of the Modern Copper and Brass Industry in Britain 1690–1750" (PhD thesis, University of Birmingham, 1985), appendix B (5).

16. Nuala Zahedieh, "Colonies, Copper and the Market for Inventive Activity in England and Wales, 1680–1730," *Economic History Review* 66, no. 3 (2013): 805–825; Nuala Zahedieh, "Technique or Demand? The Revival of the English Copper Industry ca. 1680–1730," in *Cities-Coins-Commerce: Essays Presented to Ian Blanchard on the Occasion of His Seventieth Birthday*, ed. Philipp Robinson Rössner (Stuttgart: Franz Steiner Verlag, 2012), 167–173.

17. Hilary McD. Beckles, "The 'Hub of Empire': The Caribbean and Britain in the Seventeenth Century," in *The Oxford History of the British Empire*, vol. 1, *The Origins of Empire*, ed. Nicholas Canny (Oxford: Oxford University Press, 1998), 224, table 10.1.

18. Sugar might also be prepared in cast iron pans, which were more durable. On the whole, though, copper was preferred for its superior conductivity, the ease with which pans could be cleaned, and the buoyancy of the market for scrap copper. See the discussion of the respective merits of iron and copper in Gordon Turnbull, *Letters to a Young Planter: or, Observations on the Management of a Sugar-Plantation: To Which Is Added, The Planter's Kalendar. Written on the Island of Grenada, by an Old Planter* (London: Stuart and Stevenson, 1785), 27. That copper was favored is astonishing given the price differential between copper and iron. London coppersmiths Ford & Clew would supply copper pans at sixteen pence per pound on the eve of the American Revolution (A727.1413, Forbes Papers). Iron pans cost less than two pence per pound: "Invoice of Sundries Shipped on Board the Port Morant," September 7, 1772, Slebech 11726, NLW.

19. Zahedieh, "Technique or Demand?," 169.

20. Zahedieh, "Colonies, Copper and the Market for Inventive Activity," 811, table 3.

21. Florentius Vassall to George & William Forbes, June 16, 1774, A727.1455, Forbes Papers.

22. Orders from Messrs. Long Drake & Long on July 28, 1774, and Messrs. Maitland on September 7, 1774, A727.1455, Forbes Papers.

23. Accounts 1790–1791 of the *Druid*, 39654/4, Bristol Archives. If the taches were one quarter the size of the clarifiers, a commonly assumed ratio at the time, and the intermediate coppers were sized proportionately, then a full suite of coppers would have weighed 1.6 tons. Dividing that figure by four gives the weight of an "average" copper.

24. It is also possible that more ancillary copper equipment was being used by the later eighteenth century. "There is an extraordinary great Improvement now adopted upon most Estates," it was reported from Jamaica in 1787, "in clarifying the cane liquor by flat bottomed coppers" that were used in tandem with the regular boiling coppers (John Fowler to James Stothert, April 12 and 15, 1787, James Stothert papers, William L. Clements Library, University of Michigan). This was the "apparatus for clarifying cane sugar" patented by John Reeder in October 1786: George Richardson Porter, *The Nature and Properties of Sugar Cane: With Practical Directions for the Improvement of Its Culture, and the Manufacture of Its Products* (London: Smith, Elder, & Co., 1830), 329.

25. Examination of Percival North, John Atlee, James Rogers, and Pascoe Grenfell, March 12, 1806, C/13/499/30, TNA. Work on sugar processing utensils is a prominent feature in the surviving wage books of the Battersea manufactory: C 103/68, TNA.

26. Information on these partnerships can be gained from biographical data in the *Legacies of British Slave-Ownership* database (https.//www.ucl.ac.uk/lbs/). See the entries for Samuel Boddington (1766–1843), John Wedderburn (1743–1820) of Spring Garden and St. Marylebone, and George Hibbert (1757–1837).

27. John J. McCusker, "The Business of Distilling in the Old World and the New World during the Seventeenth and Eighteenth Centuries: The Rise of a New Enterprise

and Its Connection with Colonial America," in *The Early Modern Atlantic Economy*, ed. John J. McCusker and Kenneth Morgan (Cambridge: Cambridge University Press, 2000), 186–224.

28. McCusker, "The Business of Distilling," 217–218.

29. "Names &c of Brewers, Dyers, Distillers & Sugar Bakers—London—23 May 1776," A727.1507, Forbes Papers.

30. Bryan Edwards, *The History, Civil and Commercial, of the British Colonies in the West Indies*, vol. 2 (Dublin: Luke White, 1793), 220–221.

31. McCusker, "The Business of Distilling," 193.

32. Noel Deerr, *The History of Sugar*, vol. 2 (London: Chapman & Hall Ltd., 1950), 464.

33. *An Account of the Late Application to Parliament, from the Sugar Refiners, Grocers, &c. of the Cities of London and Westminster, the Borough of Southwark, and of the City of Bristol* (London: J. Brotherton, 1753), 42.

34. Shadreck Chirikure, *Metals in Past Societies: A Global Perspective on Indigenous African Metallurgy* (Heidelberg: Springer, 2015), 78, 80–81; Caroline Robion-Brunner, "L'Afrique des métaux," in *L'Afrique ancienne de l'Acacus au Zimbabwe: 20 000 avant notre ere—xviie siècle*, ed. Francois-Xavier Fauvell (Paris: Belin, 2018), 536–539.

35. David Killick, "A Global Perspective on the Pyrotechnologies of Sub-Saharan Africa," *Azania: Archaeological Research in Africa* 51, no. 1 (2016): 77.

36. Eugenia W. Herbert, *Red Gold of Africa: Copper in Precolonial History and Culture* (Madison: University of Wisconsin Press, 2003).

37. K. G. Davies, *The Royal African Company* (London: Longmans, Green & Co., 1957), 171–172.

38. *The Origins of an Industrial Region: Robert Morris and the First Swansea Copper Works, c. 1727–1730*, ed. Louise Miskell (Newport: South Wales Record Society, 2010), 58–59; Steven Hughes, *Copperopolis: Landscapes of the Early Industrial Period in Swansea* (Aberystwyth: Royal Commission on the Ancient and Historical Monuments of Wales, 2000), 45.

39. Accounts of the *Africa*, 1774–1776, 45039, Bristol Archives.

40. Quoted in David Levine and Keith Wrightson, *The Making of an Industrial Society: Whickham 1560–1765* (Oxford: Clarendon Press, 1991), 80.

41. J. V. Beckett, *Coal and Tobacco: The Lowthers and the Economic Development of West Cumberland, 1660–1760* (Cambridge: Cambridge University Press, 1981), chapters 2 and 3; Barrie Trinder, *The Industrial Revolution in Shropshire* (Chichester: Phillimore & Co., 1981), 6–8, 53–54.

42. Hughes, *Copperopolis*, 73–74.

43. David Hussey, *Coastal and River Trade in Pre-industrial England: Bristol and Its Region, 1680–1730* (Exeter: University of Exeter Press, 2000), 96–100.

44. Rhys Jenkins, "The Reverberatory Furnace with Coal Fuel, 1612–1712," *Transactions of the Newcomen Society* 14, no. 1 (1933): 70.

45. Roger Burt, "The Transformation of the Non-ferrous Metals Industries in the Seventeenth and Eighteenth Centuries," *Economic History Review* 48, no. 1 (1995): 23–45.

46. It was the work of Sir Humphry Mackworth (1657–1727), who was bent on developing the industrial potential of his estate at Gnoll, just outside Neath. Mackworth was a member of the Company of Mine Adventurers of England, which mined lead in Cardiganshire, and the furnace at Neath, erected in the mid-1690s,

was intended to process ore that was mined at Esgair Hir, a remote location without coal or, indeed, much in the way of timber. See William P. Griffith, "Mackworth, Sir Humphry (1657–1727)," *Oxford Dictionary of National Biography* (Oxford: Oxford University Press, 2004), http://www.oxforddnb.com/view/article/17631.

47. Peter W. King, "Sir Clement Clerke and the Adoption of Coal in Metallurgy," *Transactions of the Newcomen Society* 73, no. 1 (2001): 33–52.

48. "Two Discourses on Metals by John Woodward, M.D.," Add. MS. 25095, fo. 98, British Library.

49. Jenkins, "The Reverberatory Furnace," 71; Morton, "Rise of the Modern Copper and Brass Industry," 93–99.

50. "Two Discourses on Metals by John Woodward."

51. "Two Discourses on Metals by John Woodward."

52. "An Account of the Copper Works on the Avon Three Miles above Bristol," RBO/7/123, Royal Society.

Chapter 2 • Swansea's Apprenticeship, c. 1690–1750

1. G. Hammersley, "The Effect of Technical Change in the British Copper Industry between the Sixteenth and the Eighteenth Centuries," *Journal of European Economic History* 20, no. 1 (1991): 163.

2. Joan Day, *Bristol Brass: A History of the Industry* (Newton Abbot: David & Charles, 1973).

3. John Morton, "The Rise of the Modern Copper and Brass Industry in Britain, 1690–1750" (PhD thesis, University of Birmingham, 1985), 216.

4. Clive Trott, "The Copper Industry," in *Neath and District: A Symposium*, ed. E. Jenkins (Neath: Elis Jenkins, 1974), 119–121.

5. Morton, "Rise of the Modern Copper and Brass Industry," 217.

6. Morton, "Rise of the Modern Copper and Brass Industry," 218–219.

7. Details of the running of Neath Abbey works by this partnership can be found in R. O. Roberts, "The Copper Industry of Neath and Swansea. Record of a Suit in the Court of Exchequer, 1723," *South Wales and Monmouthshire Record Society Publications* 4 (1957): 125–162.

8. William Rees, *Industry before the Industrial Revolution: Incorporating a Study of the Chartered Companies of the Society of Mines Royal and of Mineral and Battery Works*, vol. 2 (Cardiff: University of Wales Press, 1968), 502; and Morton, "Rise of the Modern Copper and Brass Industry," 325–330.

9. William P. Griffith, "Mackworth, Sir Humphry (1657–1727)," *Oxford Dictionary of National Biography* (Oxford: Oxford University Press, 2004), http://www.oxforddnb.com/view/article/17631.

10. Koji Yamamoto, "Piety, Profit and Public Service in the Financial Revolution," *English Historical Review* 126, no. 521 (2011): 806–834.

11. Rees, *Industry before the Industrial Revolution*, 521.

12. Roger Burt, "Lead Production in England and Wales, 1700–1770," *Economic History Review* 22, no. 2 (1969): 259–262.

13. Rees, *Industry before the Industrial Revolution*, 537.

14. Quoted in Trott, "The Copper Industry," 125.

15. "Minutes of the Select Committee of the Fortunate Adventurers in the Mine Adventure Company," June 15, 1703, Royal Institution of South Wales, George Grant Francis collection, 17, WGAS.

16. Letter from Robert Lydell to M. R. Fran, December 8, 1698, Company of Mine Adventurers of England correspondence, Neath Antiquarian Society collection, Gn/I 2/1, WGAS.

17. P. W. King, "Sir Clement Clerke and the Adoption of Coal in Metallurgy," *Transactions of the Newcomen Society* 73, no. 1 (2001): 44.

18. "A Calculation of the Profit and Loss in Making Copper at Neath and at Redbrook," no date, Neath Antiquarian Society collection, Gn/I 12/3, WGAS.

19. Letter from Robert Lydell to Humphry Mackworth, February 26, 1699, Neath Antiquarian Society collection, Gn I 1/2, WGAS.

20. Letter from David Pralph to Mr. Barsham, Neath, April 1749, Neath Antiquarian Society collection, Gn/I 13/2, WGAS.

21. J. U. Nef, *The Rise of the British Coal Industry*, vol. 1 (London: Cass, 1966), 19–53.

22. The figure of 25 fathoms comes from R. P. Roberts, "The History of Coal Mining in Gower 1700–1832" (MA thesis, University College of Wales, Cardiff, 1953), 101; for the northeast, see Gregory Clark and David Jacks, "Coal and the Industrial Revolution, 1700–1869," *European Review of Economic History* 11, no. 1 (April 2007): 44.

23. Roberts, "The History of Coal Mining in Gower," 40.

24. G. Thomas, "The Coal Industry," in *Neath and District. A Symposium*, ed. E. Jenkins (Neath: Elis Jenkins, 1974), 167.

25. Letter from Robert Lydell to Humphry Mackworth, February 19, 1699, Neath Antiquarian Society collection, Gn I 2/1, WGAS.

26. Nef, *Rise of the British Coal Industry*, 19.

27. For details, see Morton, "Rise of the Modern Copper and Brass Industry," 265–275.

28. William Borlase, *The Natural History of Cornwall* (Oxford: W. Jackson, 1758), 205.

29. William Pryce, *Mineralogia Cornubiensis: A Treatise on Minerals, Mines and Mining* (London: J. Phillips, 1778), 287.

30. *The Kalmeter Journal: The Journal of a Visit to Cornwall, Devon and Somerset in 1724–25 of Henric Kalmeter (1693–1750)*, ed. Justin Brooke (Truro: Twelveheads Press, 2001), 25.

31. Morton, "Rise of the Modern Copper and Brass Industry," 407–408.

32. Pocket book detailing ore trials in Cornwall by Thomas Hawkins, giving details of various mines and owners, c. 1712, Neath Antiquarian Society collection, Gn/I 1/1, WGAS.

33. For fuller details, see Borlase, *Natural History of Cornwall*, 206.

34. *Kalmeter Journal*, 34.

35. Rees, *Industry before the Industrial Revolution*, 532.

36. Morton, "Rise of the Modern Copper and Brass Industry," 279.

37. Morton, "Rise of the Modern Copper and Brass Industry," 400.

38. J. Napier, "On Copper Smelting," in *The London, Edinburgh and Dublin Philosophical Magazine and Journal of Science* 4th series, no. 4 (1852): 263–264.

39. "A Calculation of the Profit and Loss in Making Copper at Neath and at Redbrook," no date, Neath Antiquarian Society collection, Gn/1 12/3, WGAS.

40. "A Calculation of the Profit and Loss in Making Copper at Neath and at Redbrook," no date, Neath Antiquarian Society collection, Gn/1 12/3, WGAS.

41. Daniel Defoe, *A Tour through the Whole Island of Great Britain, Divided into Circuits or Journies* (London: J. M. Dent and Co., 1927), Letter VI, http://ebooks.adelaide.edu.au.

42. See, for example, "Report on Coal Works at Loughor and Penclawdd," no date, Neath Antiquarian Society collection, Gn/I 12/2, WGAS.

43. Philip Jenkins, "Tory Industrialism and Town Politics: Swansea in the Eighteenth Century," *The Historical Journal* 28, no. 1 (1985): 105–106, 117–118.

44. Stephen Hughes, *Copperopolis: Landscapes of the Early Industrial Period in Swansea* (Aberystwyth: Royal Commission on the Ancient and Historic Monuments of Wales, 2000), 26.

45. R. O. Roberts, "Dr. John Lane and the Foundation of the Non-ferrous Metal Industries in the Swansea Valley," *Gower: Journal of the Gower Society* 4 (1951): 19.

46. Joanna Martin, "Private Enterprise versus Manorial Rights: Mineral Property Disputes in Mid-Eighteenth Century Glamorgan," *Welsh History Review* 9 (January 1978): 161.

47. Day, *Bristol Brass*, 32–35.

48. Richard Dale, *The First Crash: Lessons from the South Sea Bubble* (Princeton: Princeton University Press, 2004), 137.

49. See, for example, F. V. Emery, "Fresh Light on Dr. John Lane, Co-founder of the Copper Industry at Swansea," *Gower: Journal of the Gower Society* 20 (1969): 8–13.

50. R. T. Jenkins, "Morris, Robert (d. 1768), Industrialist," *Dictionary of Welsh Biography* https://biography.wales/article/s-MORR-ROB-1768.

51. *The Origins of an Industrial Region: Robert Morris and the First Swansea Copperworks, c. 1727–1730*, ed. Louise Miskell (Newport: South Wales Record Society, 2010), 64.

52. *Origins of an Industrial Region*, 52.

53. *Origins of an Industrial Region*, 52.

54. Such were the terms on which Tom David was engaged by Robert Morris in 1728. See *Origins of an Industrial Region*, 64.

55. *Origins of an Industrial Region*, 99–100.

56. *Origins of an Industrial Region*, 63–64.

57. A definition of "wey" is included in the glossary. See also A. H. John, "Iron and Coal on a Glamorgan Estate, 1700–1740," *Economic History Review* 13, no. 1/2 (1943): 96.

58. Further details of the coal dispute between Popkin and Morris can be found in *Origins of an Industrial Region*, 43–52.

59. Pryce, *Mineralogia Cornubiensis*, 287.

60. See *Origins of an Industrial Region*, 39.

61. Edmund Newell, "'The Irremediable Evil': British Copper Smelters' Collusion and the Cornish Mining Industry, 1725–1865," in *From Family Firms to Corporate Capitalism: Essays in Business and Industrial History in Honour of Peter Mathias*, ed. K. Bruland and P. O'Brien (Oxford: Oxford University Press, 1998), 179.

62. *Origins of an Industrial Region*, 73.

63. W. H. Mulligan, "The Anatomy of Failure: Nineteenth-Century Irish Copper Mining in the Atlantic and Global Economy," in *The Irish in the Atlantic World*, ed. David Gleeson (Columbia: University of South Carolina Press, 2012), 38–52.

64. *Origins of an Industrial Region*, 70.

65. *Origins of an Industrial Region*, 59.

66. Morton, "Rise of the Modern Copper and Brass Industry," 381.

67. Llangyfelach Copper Works. The produce of copper ore and the copper wrought therein, 1733. Papers re: copper and lead smelting in Llangyfelach, 1727–1733, MS1501A, NLW.

68. Morton, "Rise of the Modern Copper and Brass Industry," 373.

69. Morton, "Rise of the Modern Copper and Brass Industry," 351.

70. *Origins of an Industrial Region*, 38–39.

71. Morton, "Rise of the Modern Copper and Brass Industry," 110–115.

72. Eveline Cruikshanks, "Lockwood, Richard (1676–1756), of College Hill, London, and Dews Hall, Maldon, Essex," *History of Parliament Online*, http://www.historyofparliamentonline.org, vol. 1715–1754.

73. A. M. Carlos and L. Neal, "Amsterdam and London as Financial Centers in the Eighteenth Century," *Financial History Review* 18, no. 1 (2011): 31.

74. Tony Coverdale, "The Ingenious Mr. Padmore: Eighteenth-Century Polymath," *International Journal for the History of Engineering and Technology* 87, no. 2 (2017): 176–189.

75. *Origins of an Industrial Region*, 87–88.

76. Greater concentration on the second half of the eighteenth century has meant that the significance of this for the Swansea District's emergence as Britain's primary center for copper smelting has not been fully explored. See, for example, H. V. Bowen, "Sinews of Trade and Empire: The Supply of Commodity Exports to the East India Company during the Late Eighteenth Century," *Economic History Review* 55, no. 3 (2002): 466–486.

77. Morton, "Rise of the Modern Copper and Brass Industry," 24; Ryuto Shimada, *The Intra-Asian Trade in Japanese Copper by the Dutch East India Company during the Eighteenth Century* (Leiden: Brill, 2005), 46–47, 66.

78. Llangyfelach Copper Works. The produce of copper ore and the copper wrought therein, 1730. Papers re: copper and lead smelting in Llangyfelach, 1727–1733, MS1501A, NLW.

79. Morton, "Rise of the Modern Copper and Brass Industry," 347.

80. Llangyfelach Copper Works. The produce of copper ore and the copper wrought therein, 1730. Papers re: copper and lead smelting in Llangyfelach, 1727–1733, MS1501A, NLW.

81. Notice re: letting of Copperworks lately used by Mr. Coster, c. 1742, Neath Antiquarian Society collection, GnI 1/1, WGAS.

82. "Agreement between Bussy Mansel of Briton Ferry, Thomas Coster of Bristol and Joseph and Samuel Percival," August 24, 1736, Local Archive Collection 122/C/1, RBA.

83. R. O. Roberts, "The White Rock Copper and Brass Works," in *Glamorgan Historian*, vol. 12, ed. R. Denning (Barry: Stewart Williams, 1981), 141.

84. Agreement between Mansel, Coster and Percival, August 24, 1736, Local Archive Collection 122/C/1, RBA.

85. Hughes, *Copperopolis*, 22.

86. Account and Memoranda Book of White Rock Copper Works, 1749–98, 12171/1, Bristol Archives.

87. Llangyfelach Copper Works. The produce of copper ore and the copper wrought therein, 1750. Papers re: copper and lead smelting in Llangyfelach, 1744–1789, MS15103-9B, NLW.

88. Robert Anthony, "'A Very Thriving Place': The Peopling of Swansea in the Eighteenth Century, *Urban History* 32, no. 1 (May 2005): 81.

89. John, "Iron and Coal on a Glamorgan Estate," 97; Defoe, *A Tour through the Whole Island of Great Britain*, vol. 2, 55.

90. *Origins of an Industrial Region*, 25–26.

Chapter 3 • Swansea's Ascendancy, 1750–1830

1. H. V. Bowen, "Sinews of Trade and Empire: The Supply of Commodity Exports to the East India Company during the Late Eighteenth Century," *Economic History Review* 55, no. 3 (2002): 466–486.

2. H. V. Bowen, *The Business of Empire: The East India Company and Imperial Britain, 1756–1833* (Cambridge: Cambridge University Press, 2006), 266.

3. R. O. Roberts, "The Smelting of Non-ferrous Metals since 1750," in *Glamorgan County History*, vol. 5, *Industrial Glamorgan*, ed. Glanmor Williams and A. H. John (Cardiff: Glamorgan County History Trust Ltd., 1980), 58.

4. For details of Chauncey Townsend's industrial enterprises in the Lower Swansea Valley, see Stephen Hughes, *Copperopolis: Landscapes of the Early Industrial Period in Swansea* (Aberystwyth: Royal Commission on the Ancient and Historic Monuments of Wales, 2000), 27, 47.

5. J. C. Symons, "The Mining and Smelting of Copper in England and Wales, 1760–1820" (MPhil thesis, Coventry University, 2003), 73, table 2.5.

6. Richard Bright, "Draft of the Particulars of the Trade of Bristol, 1788," in *The Bright-Meyler Papers: A Bristol–West India Connection, 1732–1837*, ed. Kenneth Morgan (Oxford: Oxford University Press for the British Academy, 2007), 657.

7. MS 3782/12/108/27, p. 78, Matthew Boulton papers, BA.

8. See the history of the Wimbledon mill prepared by the Wandle Industrial Museum at http://www.wandle.org/mills/wimbledonmill.pdf; survey of the manor of Kennington, 1786, Duchy of Cornwall Archive, DOC/S/108.

9. Symons, "The Mining and Smelting of Copper," 173. Table 2.5 compares the metallic content of the ores raised in Cornwall and Anglesey.

10. J. R. Harris, *The Copper King: A Biography of Thomas Williams of Llanidan* (Liverpool: Liverpool University Press, 1964), 39.

11. Harris, *The Copper King*, 66.

12. Richard Crawshay to James Cockshutt, March 25, 1789, D1.182, Gwent Archives.

13. See Symons, "The Mining and Smelting of Copper," 72.

14. William Pryce, *Mineralogia Cornubiensis: A Treatise on Minerals, Mines and Mining* (London: J. Phillips, 1778), 288.

15. Thomas Brown to William Grenfell, August 6, 1829, BMSS 12657, BUA.

16. Thomas Brown to William Grenfell, June 28, 1830, BMSS 12747, BUA.

17. Thomas Brown to William Grenfell, July 7, 1830, BMSS 12751, BUA.

18. See, for example, A. Bielenberg, "Industrial Growth in Ireland, c. 1790–1910" (PhD thesis, London School of Economics, 1994), 199.

19. Memorandum of weekly purchases of ores in Cornwall and Swansea by Owen Williams, April 23, 1829–March 21, 1832, BMSS 12259, BUA.

20. E. Donovan, *Descriptive Excursions through South Wales and Monmouthshire, in the Year 1804, and the Four Preceding Summers*, vol. 2 (London: Rivingtons, 1805), 52.

21. J. H. Vivian, "An Account of the Process of Smelting Copper as Conducted at the Hafod Copper Works, near Swansea," *Annals of Philosophy*, new series, 5 (1823): 116.

22. Randolph Cock, "'The Finest Invention in the World': The Royal Navy's Early Trials of Copper Sheathing, 1708–1770," *Mariner's Mirror* 87, no. 4 (2001): 446–459; J. R. Harris, "Copper and Shipping in the Eighteenth Century," *Economic History Review* 19, no. 3 (1966): 550–568.

23. Peter M. Solar and Klas Rönnbäck, "Copper Sheathing and the British Slave Trade," *Economic History Review* 68, no. 3 (2015): 807.

24. MS 3782/12/108/27, p. 70, Matthew Boulton papers, BA.

25. Figure calculated from *Report of the Committee Appointed to Enquire into the State of the Copper Mines and Copper Trade of Great Britain*, 1799, appendix 38.

26. See, for example, E. Newell, "'The Irremediable Evil': British Copper Smelters' Collusion and the Cornish Mining Industry, 1725–1865," in *From Family Firms to Corporate Capitalism: Essays in Business and Industrial History in Honour of Peter Mathias*, ed. K. Bruland and P. O'Brien (Oxford: Oxford University Press, 1998), 182.

27. *Report of the Committee Appointed to Enquire into the State of the Copper Mines and Copper Trade of Great Britain*, 1799. See minutes of evidence.

28. Harris, *The Copper King*, 115–138.

29. Pryce, *Mineralogia Cornubiensis*, 288.

30. Newell, "'Irremediable Evil,'" 179–183.

31. R. O. Roberts, "Enterprise and Capital for Non-ferrous Metal Smelting in Glamorgan, 1694–1924," *Morgannwg. Transactions of the Glamorgan Local History Society* 23 (1979): 59.

32. For details, see Harris, *The Copper King*, 120–123.

33. P. Watts-Russell, "Coal, Copper, Copperopolis . . . A Tale of Smelting Copper at Middle and Upper Bank Works, Swansea," *Journal of the Trevithick Society*, no. 46 (2019): 14.

34. E. Newell, "Grenfell Family (per. c. 1785–1879), Copper Smelters," *Oxford Dictionary of National Biography* (Oxford University Press, 2004), http://www.oxforddnb.com/view/article/61181.

35. Reports of Upper Bank copper works, BMSS 12277, BUA.

36. See, for example, Greenfield Report, August 29, 1829, BMSS 12282, BUA.

37. Andrew Ure, *A Dictionary of Arts, Manufactures and Mines: Containing a Clear Exposition of Their Principles and Practice* (New York: D. Appleton and Co., 1842), 223–224.

38. See, for example, Stock of Copper at the Liverpool Warehouse, November 5, 1829, BMSS, 12283, BUA.

39. John Vivian to J. H. Vivian, January 30, 1810, Vivian A501, NLW.

40. Hughes, *Copperopolis*, 30–34.

41. John Vivian to J. H. Vivian, February 2, 1814, Vivian A657, NLW.

42. Ure, *Dictionary of Arts, Manufactures and Mines*, 330.

43. Report of Hafod Smelting Works and Forest Mills, week ending November 20, 1828, Vivian E113, NLW.

44. Vivian, "An Account of the Process of Smelting Copper," 123.

45. John Vivian to J. H. Vivian, May 8, 1811, Vivian A594, NLW.

46. Memorandum Book on Copper Trade, MS15111B, NLW.

47. E. Newell, "The British Copper Ore Market in the Nineteenth Century with Particular Reference to Cornwall and Swansea" (PhD thesis, University of Oxford, 1988), 19.

48. R. Martello, "Paul Revere's Last Ride: The Road to Rolling Copper," *Journal of the Early Republic* 20, no. 2 (Summer 2000): 226–227, 236.

49. John Vivian to J. H. Vivian, July 4, 1813, Vivian A641, NLW.

50. John Vivian to J. H. Vivian, July 12, 1823, Vivian A849, NLW.

51. A. Jamieson, *A Dictionary of Mechanical Science, Arts, Manufactures and Miscellaneous Knowledge, vol. 1* (London: Henry Fisher, Son & Co., 1829), 195.

52. See, for example, *North Wales Gazette*, October 4, 1824.

53. H. Davy, "Additional Experiments and Observations on the Application of Electrical Combinations to the Preservation of the Copper Sheathing of Ships . . . ," *Philosophical Transactions* 114 (December 1824).

54. "Mushet's Patent Sheathing Copper," *The Oriental Herald Advertiser*, September 1825.

55. See, for example, John Vivian to J. H. Vivian, September 7, 1823, Vivian A852, NLW. Regulus, referred to by John Vivian as "regule" in this letter, was a purer form of copper separated from the ore after the first stage of the smelting process.

56. *The Cambrian*, October 25, 1834.

57. The full name of the company was Pitt, Anderson, Birch and Company. See Joan Day, *Bristol Brass. A History of the Industry* (Newton Abbot: David & Charles, 1973), 131.

58. Ure, *A Dictionary of Arts, Manufactures and Mines*, 330.

59. John Vivian to J. H. Vivian, November 14, 1809, Vivian A481, NLW.

60. Bielenberg, "Industrial Growth in Ireland," 22–28.

61. Agreement for the supply of copper, October 25, 1822, Vivian B78, NLW.

62. See, for example, details of an order from O'Connor, John Vivian to J. H. Vivian, May 17, 1813, Vivian A629, NLW.

63. George Unwin, *Observations upon the Export Trade of Tin and Copper to India with Reference to the Expected Renewal of the Honourable East India Company's Charter* (Truro: T. Flindell, 1811), 12–13.

64. Bowen, "Sinews of Trade," 474.

65. John Vivian to J. H. Vivian, August 1, 1809, Vivian A476, NLW.

66. Bowen, "Sinews of Trade," 471.

67. John Vivian to J. H. Vivian, June 9, 1822, Vivian A817, NLW.

68. Bowen, "Sinews of Trade," 470–471.

69. An Account of the Quantities of Copper Exported from Great Britain in the Year Ending 5 January 1820, Vivian E12, NLW.

70. Notes on the Sale of Copper in India, Vivian E10, NLW; John Vivian to J. H. Vivian, May 22, 1812, Vivian A617, NLW.

71. Vivian, "An Account of the Process of Smelting Copper," 122.

72. *Michael Faraday in Wales: Including Faraday's Journal of His Tour through Wales in 1819*, ed. D. Tomas (Denbigh: Gwasg Gee, 1972), 36. See also John Scoffern, William Truran, William Clay, Robert Oxland, William Fairbairn, W. C. Aitkin, and William Vose Pickett, *The Useful Metals and Their Alloys, including Mining Ventilation, Mining Jurisprudence, and Metallic Chemistry Employed in the Conversion of Iron, Copper, Tin, Zinc, Antimony and Lead Ores; with Their Application to the Industrial Arts* (London: Houlston and Wright, 1866), 551.

73. A. Webster, "The Political Economy of Trade Liberalization: The East India Company Charter Act of 1813," *Economic History Review*, new series, 43, no. 3 (1990): 404–419.

74. John Vivian to J. H. Vivian, February 16, 1813, Vivian A628, NLW.

75. Population figures from John Williams, *Digest of Welsh Historical Statistics*, vol. 1 (Cardiff: Welsh Office, 1985), 64–65.

76. W. H. Jones, *History of the Port of Swansea* (Carmarthen: Spurrell & Son, 1922), appendix.

77. For details, see Louise Miskell, *Intelligent Town. An Urban History of Swansea, c. 1780–1855* (Cardiff: University of Wales Press, 2006), 41–69.

78. *The New Swansea Guide: Containing a Particular Description of the Town and Its Vicinity* (London: Longman, 1823), 22.

79. See, for example, G. Nicholson, *The Cambrian Traveller's Guide and Pocket Companion* (Stourport: George Nicholson, 1808); J. Feltham, *A Guide to the Watering and Sea-Bathing Places* (London: Richard Phillips, 1806), 396.

Chapter 4 • Global Swansea, 1830–1843

1. Edmund Newell, "'Copperopolis': The Rise and Fall of the Copper Industry in the Swansea District, 1826–1921," *Business History* 32, no. 3 (1990): 75–97, especially 78–80.

2. Frédéric Le Play, *Description des procédés métallurgiques employés dans le Pays de Galles pour la fabrication du cuivre* (Paris: Carilian-Goeury et Von Dalmont, 1848), 6–7.

3. These developments are summarized in Chris Evans and Olivia Saunders, "A World of Copper: Globalizing the Industrial Revolution, 1830–1870," *Journal of Global History* 10, no. 1 (2015): 3–26.

4. Le Play, *Description*, 387: "fonderie centrale des minerais des deux océans."

5. Richard Hussey Vivian to J. H. Vivian, January 5, 1825, Vivian A1017, NLW.

6. Quoted in David R. Fisher, "Vivian, Sir Richard Hussey (1775–1852), of Beechwood House, nr. Lyndhurst, Hants.," in *The History of Parliament: The House of Commons 1820–1832*, ed. D. R. Fisher (2009), http://www.historyofparliamentonline.org/volume/1820-1832/member/vivian-sir-richard-1775-1842. Richard Hussey Vivian, who was MP for Truro and a champion of Cornish mining, spoke up in Parliament against the lowering of duty on foreign ores.

7. Charles Pascoe Grenfell private ledger 1828–1835, D/GR 9/1, CBS. Several South American mining companies are listed in the index to this volume; the relevant pages have been torn out.

8. Edmund Newell, "The British Copper Ore Trade in the Nineteenth Century, with Particular Reference to Cornwall and Swansea" (DPhil thesis, University of Oxford, 1988), 82.

9. Copper. Accounts relating to the import of copper, copper ore, brass, and copper manufactures, 1847, BPP, 637, table 2.

10. María Elena Díaz, *The Virgin, the King, and the Royal Slaves of El Cobre: Negotiating Freedom in Colonial Cuba, 1670–1780* (Stanford: Stanford University Press, 2000).

11. John Hardy Jr. (1806–1842) was the son of a London-based West India merchant of the same name. I am grateful to Stefan Krzeczunowicz of Toronto for sharing his research on the family with me.

12. Hardy's report on the commerce of Santiago de Cuba, March 3, 1833, enclosed with John Hardy Jr. to W. S. McLeay, February 23, 1833, FO 453/1, TNA.

13. David Turnbull, *Travels in the West. Cuba; with Notices of Porto Rico, and the Slave Trade* (London: Longman, Orme, Brown, Green, and Longmans, 1840), 10.

14. The British partners were Hardy and his father, John Hardy Sr. The local participants were Prudencio Casamayor (1763–1842), a French-born refugee from Saint-Domingue, Antonio San Emeterio, and José Touson. See the prospectus of the Cobre Company (1835), 2, enclosed with John Hardy Jr. to Lord Palmerston, December 27, 1836, FO 84/201, TNA; Inés Roldán de Montaud, "El ciclo cubano del cobre en el siglo XIX, 1830–1868," *Bolétin Geológico y Minero* 119, no. 3 (2008): 362–363. They were joined later by Joaquín de Arrieta of Havana, who wielded "a vast deal of influence" and had been indispensable in obtaining a mining license: Turnbull, *Travels in the West*, 11.

15. Prospectus of the Cobre Company, 4.

16. Registration documents of December 11, 1866, reciting an earlier indenture of July 13, 1835, BT 31/1310/3371, TNA.

17. This development is dealt with more fully in Chris Evans, "El Cobre: Cuban Ore and the Globalization of Swansea Copper, 1830–1870," *Welsh History Review* 27, no. 1 (2014): 112–131.

18. See the biographical notes accompanying the family papers at NLW (http://www.archiveswales.org.uk/anw/get_collection.php?coll_id=20206&inst_id=1&term=Goring-Thomas%20family%2C%20|%20of%20Llannon%20|%20Archives).

19. Materials in the Baring Archive give an impression of the firm's range: HC3.51.9 (statement by J. Horsley Palmer on Georges Wildes & Co., March 23, 1837), HC3.51.12 (correspondence between Wildes & Co. and Barings 1837–1838), and HC3.52.3 (correspondence and accounts concerning settlement of the affairs of Wildes & Co., 1837–1847).

20. Sir John Pirie (he was knighted after serving as London's lord mayor in 1841) was obituarized in *The Gentleman's Magazine and Historical Revie*w 189 (1851): 551–552, although the focus there was on his civic accomplishments rather than his "great commercial importance as a merchant and shipowner."

21. Geoffrey Alderman, "Goldsmid, Sir Isaac Lyon, first baronet (1778–1859)," *Oxford Dictionary of National Biography* (Oxford: Oxford University Press, 2004),

http://www.oxforddnb.com/view/article/10920; Martin Daunton, "Thompson, William (1793–1854)," *Oxford Dictionary of National Biography* (Oxford: Oxford University Press, 2004), http://www.oxforddnb.com/view/article/40810.

22. Michael Williams is noticed in the *Oxford Dictionary of National Biography* in the entry for John Charles Williams, his grandson: Garry Tregidga, "Williams, John Charles (1861–1939)," *Oxford Dictionary of National Biography* (Oxford: Oxford University Press, 2009), http://www.oxforddnb.com/view/article/98421.

23. R. O. Roberts, "The Smelting of Non-ferrous Metals since 1750," in *Glamorgan County History*, vol. 5, *Industrial Glamorgan from 1700 to 1970*, ed. A. H. John and Glanmor Williams (Cardiff: Glamorgan County History Trust, 1980), 59.

24. *Morning Chronicle*, March 19, 1838.

25. HJ1, RIC.

26. Diary of James Whitburn, AD 1341, Cornwall Record Office.

27. *Royal Copper Mines of Cobre Association, Second Report, October 25th, 1836.* This is shelved in the British Library at 1890.e.1.105.

28. Alfred Jenkin to William Leckie, July 12, 1836, HJ1/17, RIC.

29. *Morning Chronicle*, March 19, 1838.

30. William Nicholl, a miner from Illogan, near Redruth, agreed to serve the Santiago Company for three years according to the pro forma contract he signed on December 10, 1842, which is enclosed with Charles Clarke to Lord Aberdeen, August 13, 1843, FO 72/634, TNA.

31. Turnbull, *Travels in the West*, 9.

32. Alfred Jenkin to James Poingdestre, August 9, 1837, HJ1/17, RIC.

33. Quoted in Keith Strange, *Merthyr Tydfil, Iron Metropolis: Life in a Welsh Industrial Town* (Stroud: The History Press, 2005), 170. The Santiago Company also recruited in northeast Wales: see James Treweck to Charles Clarke, April 23, 1843, FO 72/634, TNA.

34. Charles Clarke to Joseph Crawford, April 27, 1843, FO 72/634, TNA.

35. "Summary of the Distribution of the Operatives Employed at the Royal Consolidated Cobre Mines," enclosed with John Hardy Jr. to Lord Palmerston, December 27, 1836, FO 84/201, TNA.

36. For fuller detail of this process, see Chris Evans, "Carabalíes y culíes en El Cobre: esclavos africanos y trabajadores chinos al servicio del cobre para Swansea, siglo XIX," *Revista de Historia Social y de las Mentalidades* 21, no. 1 (2017): 181–209.

37. This figure is drawn from *Voyages: The Trans-Atlantic Slave Trade Database.*

38. John Hardy Jr. to Lord Palmerston, December 27, 1836, FO 84/201, TNA.

39. AD 1341, Cornwall Record Office.

40. Vicente González Loscertales and Inés Roldán de Montaud, "La minería del Cobre en Cuba. Su organización, problemas administrativos y repercusiones sociales (1828–1849)," *Revista de Indias* 40 (1980): 275.

41. Chris Evans, "Brazilian Gold, Cuban Copper and the Final Frontier of British Anti-Slavery," *Slavery & Abolition* 34, no. 1 (2013): 122 and 131, n. 18.

42. *The British and Foreign Anti-Slavery Reporter*, September 22, 1841.

43. Evans, "Brazilian Gold, Cuban Copper."

44. BPP 1849 (1128), Class B: Correspondence with British ministers and agents in foreign countries, and with foreign ministers in England, relating to the slave trade, from April 1, 1848, to March 31, 1849: 320.

45. Gonzalez Loscertales and Roldán de Montaud, "La minería del Cobre en Cuba," 277.

46. Charles William Centner, "Great Britain and Chilean Mining, 1830–1914," *Economic History Review* 12, nos. 1–2 (1942): 78–86.

47. Manuel Llorca-Jaña, "Exportaciones chilenas de cobre a Gales durante el siglo XIX: su impacto en las economías chilena y galesa," *Revista de Historia Social y de las Mentalidades* 21, no. 1 (2017): 27–62.

48. Luis Valenzuela, "The Chilean Copper-Smelting Industry in the Mid-Nineteenth Century: Phases of Expansion and Stagnation, 1834–1858," *Journal of Latin American Studies* 24, no. 3 (1992): 507–550. See table 1 at 513.

49. John Mayo, "Commerce, Credit and Control in Chilean Copper Mining before 1880," in *Miners and Mining in the Americas*, ed. Thomas Greaves and William Culver (Manchester: Manchester University Press, 1985), 29–46 (at 35–36).

50. Louise Miskell, "From Copperopolis to Coquimbo: International Knowledge Networks in the Copper Industry of the 1820s," *Welsh History Review* 27, no. 1 (2014): 92–111; Claudio Veliz, "Egaña, Lambert and the Chilean Mining Associations of 1825," *Hispanic American Historical Review* 55, no. 4 (1975): 637–663; Henry English, *A General Guide to the Companies Formed for Working Foreign Mines* (London: Boosey & Sons, 1825), 19–20.

51. Quoted in Mayo, "Commerce, Credit and Control," 33.

52. Charles Darwin, *Journal of Researches into the Natural History and Geology of the Countries Visited during the Voyage round the World of H.M.S.* Beagle (London: John Murray, 1913), 277.

53. Jules Ginswick, ed., *Labour and the Poor in England and Wales, 1849–1851: The Letters to* The Morning Chronicle *from the Correspondents in the Manufacturing and Mining Districts, the Towns of Liverpool and Birmingham, and the Rural Districts*, vol. 3, *The Mining and Manufacturing Districts of South Wales and North Wales* (London: Frank Cass & Co., 1983), 184.

54. John Scoffern, William Truran, William Clay, Robert Oxland, William Fairbairn, W. C. Aitkin, and William Vose Pickett, *The Useful Metals and Their Alloys, including Mining Ventilation, Mining Jurisprudence, and Metallic Chemistry Employed in the Conversion of Iron, Copper, Tin, Zinc, Antimony and Lead Ores; with Their Application to the Industrial Arts* (London: Houlston & Wright, 1866), 551.

55. Ginswick, *Labour and the Poor*, 187, 198.

56. Ginswick, *Labour and the Poor*, 195.

57. *Morning Chronicle*, August 11, 1843.

58. Ginswick, *Labour and the Poor*, 192.

59. Ronald Rees, *King Copper: South Wales and the Copper Trade 1584–1895* (Cardiff: University of Wales Press, 2000), 56.

60. Le Play, *Description*, 162.

61. Richard Warner, *A Second Walk through Wales . . . in August and September 1798*, 2nd edition (London: G. G. and J. Robinson, 1800), 88.

62. Samuel Rowland Fisher, journal of English travels 1783–1784, October 18, 1783, Collection 2019, Historical Society of Pennsylvania. This was the Ravenhead works in Lancashire where ores from Parys Mountain were smelted.

63. Reverend J. Evans, *Letters Written during a Tour through South Wales in the Year 1803* (London: C. & R. Baldwin, 1804), 147.

64. See Rees, *King Copper*, chapters 4 to 7, for extensive coverage of the copper smoke controversy.

65. Roberts, "The Smelting of Non-ferrous Metals since 1750," 71–72.

66. *The Cambrian*, August 12, 1843.

67. *The Cambrian*, August 12, 1843.

68. Ginswick, *Labour and the Poor*, 185.

69. Ronald Rees, for example, reproduces the *Morning Chronicle*'s verdict verbatim (but without attribution): strikes were "a very rare occurrence indeed." He goes on to argue that workers knew lives of "relative comfort and security" (which is arguable) and responded with "absolute loyalty to the company" that employed them (which is false). Rees, *King Copper*, 57–58.

70. Price-fixing was endemic in the iron trade, so often seen as the polar opposite of copper: Chris Evans, "The Corporate Culture of the British Iron Industry 1650–1830," in *The Social Organization of the European Iron Industry 1600–1900*, ed. Göran Rydén (Stockholm: Jernkontoret, 1997), 121–146.

71. For seniority systems, see Alastair J. Reid, *Social Classes and Social Relations in Britain, 1850–1914* (Cambridge: Cambridge University Press, 1995), 33–35; Takao Matsumura, *The Labour Aristocracy Revisited: The Victorian Flint Glass Makers 1850–1880* (Manchester: Manchester University Press, 1983).

72. Michael Z. Brooke, *Le Play: Engineer and Social Scientist* (London: Longman, 1970), 41–42.

73. Le Play, *Description*, 117–118, 175.

74. Le Play, *Description*, 238.

75. Le Play, *Description*, 318.

76. Agreement between Daniell & Co. and William Harry, August 22, 1820, Nevill 676, NLW.

77. Thomas Brown to Owen Williams, May 30, 1829, 12639, BUA. Brown's departure from Middle Bank provided an opening for Thomas Rees, an Upper Bank refiner who moved in the opposite direction to take over Brown's old role of assistant superintendent at Middle Bank: Thomas Brown to William Grenfell, July 18, 1829, 12651, BUA.

78. "Journal de voyage dans le pays de Gales et Cornwall par Ms. Ellicott et Lacour," 23, J 1864, Report 307, École des Mines, Paris.

79. Joan Day, *Bristol Brass: A History of the Industry* (Newton Abbot: David & Charles, 1973), 56.

80. Thomas Brown to William Grenfell, July 27, 1829, 12653, BUA.

81. Le Play, *Description*, 89.

82. Quoted in Roberts, "The Smelting of Non-ferrous Metals since 1750," 66.

83. P. Richards and M. Small, *Hafod and the Lower Swansea Valley: Understanding Urban Character* (Cadw: Welsh Government, 2016), 27.

84. Account and Memoranda book of White Rock Copper Works, 1749–1798, 12171/1, p. 41, Bristol Archives.

85. Thomas Brown to William Grenfell, April 10, 1830, 12729, BUA.

86. Ginswick, *Labour and the Poor*, 192.

87. Le Play, *Description*, 88, 89.

88. Le Play, *Description*, 115–116.

89. Plan of Middle Bank and Upper Bank, LAC/45/D.20/24, RBA.

90. Ginswick, *Labour and the Poor*, 188.

91. MS 15113B, NLW, 210–211.

92. Weekly reports, July 4 to December 26, 1829, 12277, BUA.

93. *The Origins of an Industrial Region: Robert Morris and the First Swansea Copper Works, c. 1727–1730*, ed. Louise Miskell (Cardiff: South Wales Record Society, 2010), 40–41.

94. MS 15113B, NLW, 22.

95. R. R. Toomey, "Vivian and Sons, 1809–1924: A Study of the Firm in the Copper and Related Industries" (PhD thesis, University of Wales, 1979), 128–129.

96. David Hussey, *Coastal and River Trade in Pre-industrial England: Bristol and Its Region 1680–1730* (Exeter: Exeter University Press, 2000), 51, tables 2.7 and 2.8.

97. Boulton's notebook "Copper 1780," MS 3782/12/108/27, BA.

98. John Henry Vivian, "An Account of the Process of Smelting Copper as Conducted at the Hafod Copper Works, Near Swansea," *Annals of Philosophy*, new series, 5 (1823): 123. Le Play, writing in the 1840s, reckoned that barks operating in the Bristol Channel were usually of between 100 and 150 tons, with a crew of eight: *Description*, 57.

99. Ginswick, *Labour and the Poor*, 182.

100. Ginswick, *Labour and the Poor*, 183.

101. Vivian, "An Account," 115.

102. James Napier, "On Copper Smelting," *The London, Edinburgh and Dublin Philosophical Magazine and Journal of Science* fourth series 4 (1852): 45–59, 192–201, 262–271, 345–355, 453–465; fourth series, 5 (1853): 30–39, 175–184, 345–354, 486–493.

103. Notes on copper smelting on the Swansea and Neath rivers, c. 1808, B29, Vivian MSS, NLW.

104. John Morton, "The Rise of the Modern Copper and Brass Industry in Britain 1690–1750" (PhD thesis, University of Birmingham, 1985), 56.

105. Loose leaf in Boulton's notebook "Copper 1780," MS 3782/12/108/27, BA.

106. Napier, "On Copper Smelting," 346.

107. MS 3782/12/108/27, BA, 93.

108. They were not alone in this. An escalation of workloads was a feature of many coal-burning trades in the early industrial age. See Chris Evans, "Work and Workloads during Industrialization: The Experience of Forgemen in the British Iron Industry 1750–1850," *International Review of Social History* 44, no. 2 (1999): 197–215, for heightened workloads in the iron industry.

109. Le Play, *Description*, 309. For many, in fact, the workweek began on Sunday evenings. Le Play (199) reported that one smelter in four was called in to heap up the fires at a set of adjacent furnaces so they would be at a white heat on Monday mornings.

110. Le Play, *Description*, 246.

111. Diary of John Place, quoted in George Grant-Francis, *The Smelting of Copper in the Swansea District of South Wales, from the Time of Elizabeth to the Present Day* (London: H. Sotheran & Co., 1881), 78.

112. Ginswick, *Labour and the Poor*, 192.

113. Le Play, *Description*, 147–148.

114. Le Play, *Description*, 117–118, 147.

115. Le Play, *Description*, 175.

116. Le Play, *Description*, 185.

117. Le Play, *Description*, 173–174.

118. Le Play, *Description*, 186.

119. Le Play, *Description*, 198.

120. Le Play, *Description*, 223.

121. Le Play, *Description*, 238.

122. *The Times*, August 8, 1843.

123. Diary and letterbook of Charles Nevill Sr., entry for August 19, 1797, Nevill 1-56/7, NLW.

124. Diary of John Place, December 18, 1797, quoted in Grant-Francis, *The Smelting of Copper*, 78.

125. Diary of John Place, January 11 and 18, 1798, quoted in Grant-Francis, *The Smelting of Copper*, 78.

126. Quoted in Toomey, "Vivian and Sons," 160.

127. MS 15113B, NLW, 121.

128. Thomas Brown to William Grenfell, June 28, 1830, 12747, BUA.

129. Thomas Brown to William Grenfell, July 7, 1830, 12751, BUA.

130. Douglas Hay and Paul Craven, "Introduction," in *Masters, Servants, and Magistrates in Britain and the Empire, 1562–1955*, ed. Douglas Hay and Paul Craven (Chapel Hill: University of North Carolina Press, 2004), 1–2.

131. See Chris Evans, *The Labyrinth of Flames: Work and Social Conflict in Early Industrial Merthyr Tydfil* (Cardiff: University of Wales Press, 1993), 89–107, for the use of this and other punitive legislation by Welsh ironmasters.

132. James Moher, "From Suppression to Containment: Roots of Trade Union Law to 1825," in *British Trade Unionism 1750–1850: The Formative Years*, ed. John Rule (London: Longman, 1988), 74–97.

133. Combination Act, 1799, 39 Geo. III c. 81.

134. Combination Act, 1800, 39 & 40 Geo. III c. 106.

135. Quarter sessions roll Michaelmas 1820, Q/S/R 1820 D, GA.

136. G. N. Grenfell to the Dowlais Iron Co., August 29, 1820, DG/A/1/69, f. 471, GA; *The Cambrian*, September 2, 1820, 3.

137. John Vivian to John Henry Vivian, August 29, 1820, A773, Vivian papers, NLW.

138. Samuel Bordell to Guest & Co., September 23, 1820, DG/A/1/68, 109–110, GA. Bordell named eight "Ore Calciner Men," six "Slag Furnace Men," eight "Metal Furnace Men," and fifteen "Ore Furnace Men."

139. John Vivian to J. H. Vivian, August 29, 1820, A773, Vivian papers, NLW. It is not clear what act is meant by the "20th and 21st Geo. 2d." It may refer to the Regulation of Servants and Apprentices Act of 1746.

140. Ginswick, *Labour and the Poor*, 184–185.

141. MS 15113B, NLW, 14. It was an abiding concern of the employers. The manager of Margam complained in 1910 that the restriction of output was as bad as he had known it in forty years' acquaintance with the industry. Toomey, "Vivian and Sons," 166–167.

142. MS 15113B, NLW, 20.

143. Christopher J. Schmitz, *World Non-ferrous Metal Production and Prices, 1700–1976* (London: Frank Cass, 1979), 64.

144. Schmitz, *World Non-ferrous Metal Production*, 268–269.

145. The English Copper Company was a long-established corporate entity, dating back to the first flush of the British copper sector in the 1690s, which folded at the end of the 1830s. An entirely new company took over its assets and assumed its corporate identity, operating copper works at Forest in the Lower Swansea Valley and Cwmavon on the eastern edge of the Swansea District. See R. O. Roberts, "Enterprise and Capital for Non-ferrous Metal Smelting in Glamorgan, 1694–1924," *Morgannwg. Transactions of the Glamorgan Local History Society* 23 (1979): 61.

146. *The Cambrian*, August 12, 1843.

147. Adam Godfrey, "'We would rather starve idle than working hard': The Swansea Copperworkers' Strike of 1843," *Llafur* 11, no. 4 (2015): 9.

148. "And they blessed Rebekah, and said unto her, Thou *art* our sister, be thou *the mother* of thousands of millions, and let thy seed possess the gate of those which hate them" (Genesis, 24:40).

149. David J. V. Jones, *Rebecca's Children: A Study of Rural Society, Crime, and Protest* (Oxford: Clarendon Press, 1989).

150. *The Times*, August 8, 1843.

151. Jones, *Rebecca's Children*, 125.

152. Evans, "El Cobre"; Jennifer Protheroe-Jones, email, January 20, 2010.

153. *Morning Chronicle*, August 3, 1843.

154. *Morning Chronicle*, August 3, 1843.

155. *The Times*, August 8, 1843.

156. *Morning Chronicle*, August 11, 1843. In fact, a Welsh farm laborer would be very fortunate to taste bread, bacon, or cabbage, according to an unflattering account in the *Morning Chronicle* on August 3, 1843: "His food consists of the very coarsest oaten cakes called 'barra cyrch,' which to an Englishman would be most nauseous. Broth of the thinnest possible description. These ingredients, with a trifling supply of potatoes, form the labourer's daily food."

157. *The Times*, August 8, 1843.

158. Ginswick, *Labour and the Poor*, 192.

159. *The Times*, August 8, 1843.

160. Reprinted in the *Morning Chronicle*, August 11, 1843.

161. *The Cambrian*, August 12, 1843.

162. *The Times*, August 8, 1843.

163. *The Cambrian*, August 26, 1843.

164. *Morning Chronicle*, August 11, 1843.

165. *The Cambrian*, August 26, 1843.

166. *The Cambrian*, September 9, 1843.

167. *The Cambrian*, September 9, 1843.

168. *The Cambrian*, September 9, 1843.

169. *The Cambrian*, September 16, 1843.

170. Edmund Newell, "'The Irremediable Evil': British Copper Smelters' Collusion and the Cornish Mining Industry, 1725–1865," in *From Family Firms to Corporate*

Capitalism: Essays in Business and Industrial History in Honour of Peter Mathias, ed. Kristine Bruland and Patrick O'Brien (Oxford: Clarendon Press, 1998), 170–198.

171. *The Cambrian*, September 30, 1843.

Chapter 5 • Global Fragmentation, 1843–c. 1870

1. Luis Valenzuela, "The Chilean Copper Smelting Industry in the Mid-Nineteenth Century: Phases of Expansion and Stagnation, 1834–58," *Journal of Latin American Studies* 24, no. 3 (1992): 516–517. Coal mining around the southern port of Concepción began at this time, although it would be the 1860s before Chilean production matched foreign imports. See also Luis Ortega, "The First Four Decades of the Chilean Coal Mining Industry, 1840–1879," *Journal of Latin American Studies* 14, no. 1 (1982): 1–32 (at 2).

2. "The Memorial of the Directors of the Chamber of Commerce and Manufactures at Manchester," BPP, 1847–1848 (186), 4.

3. Valenzuela, "The Chilean Copper Smelting Industry," 521.

4. Valenzuela, "The Chilean Copper Smelting Industry," 521.

5. R. J. Nevill of the Llanelly Copper Company in August 1846, quoted in Valenzuela, "The Chilean Copper Smelting Industry," 522–523.

6. In reality, the legislative changes of 1843 registered a political economy that had already come into being; they did not inaugurate it. The prohibition on machinery exports had been flouted for decades: see Gillian Cookson, *The Age of Machinery: Engineering the Industrial Revolution, 1770–1850* (Woodbridge: The Boydell Press, 2018), 252. Besides, the export of machinery was of little importance in the copper industry, where the key material ingredient was the firebrick. As for the emigration of skilled workers, the success of the Cobre Mining Association in sending Cornish miners to Cuba in the 1830s suggests that legal restrictions had little effect.

7. Philip Payton, *The Cornish Overseas* (Fowey: Alexander Associates, 1999), 166–201.

8. James Napier, "On Copper Smelting," *The London, Edinburgh and Dublin Philosophical Magazine and Journal of Science*, fourth series, 4 (1852): 49. Napier specified seven different grades of Burra ore, of which the lowliest was 16 percent copper. The four richest grades ranged from 37 to 44 percent.

9. *South Australian*, March 14, 1848.

10. "The Burra Burra Copper-Mine," *Illustrated London News*, December 2, 1848.

11. Jason Shute, *Henry Ayers: The Man Who Became a Rock* (London: I. B. Tauris, 2011), 23.

12. "Yatala Smelting Works," reprinted in the *Sydney Morning Herald*, April 16, 1849.

13. "The Yatala Smelting Works," *South Australian*, September 26, 1848, reprinted in *The Maitland Mercury and Hunter River Advertiser*, November 11, 1848.

14. A. G. Banks, *H. W. Schneider of Barrow and Bowness* (Kendal: Titus Wilson, 1984); Aidan C. J. Jones, "Schneider, Henry William (1817–1887)," *Oxford Dictionary of National Biography* (Oxford: Oxford University Press, 2004), http://www.oxforddnb.com/view/article/50511.

15. G. J. Drew, *The Leyshon Joneses: Father and Son Welsh Smeltermen Who Dominated the South Australian Smelting Industry from 1848–1877* (Adelaide: Greg Drew, 2018), 13.

16. R. O. Roberts, "The Bank of England, the Company of Copper Miners and the Cwmavon Works 1847–52," *Welsh History Review* 4, no. 3 (1969): 219–234; Roberts, "Enterprise and Capital," 61.

17. *The Register*, October 4, 1848, quoted in Drew, *The Leyshon Joneses*, 13.

18. *The Register*, July 8, 1851, quoted in Drew, *The Leyshon Joneses*, 18.

19. Mel Davies, "Balanced Costs: Inland Copper Smelting Location and Fuel in South Australia 1848–76: Were They So Naive?," *University of Western Australia, Department of Economics, Working Paper 05-25* (2005), 1–17.

20. "The Burwood Copper-Works," *Illustrated London News*, February 11, 1854.

21. Philip Payton, *Making Moonta: The Invention of "Australia's Little Cornwall"* (Exeter: Exeter University Press, 2007).

22. Drew, *The Leyshon Joneses*, 28.

23. J. B. Austin, *The Mines of South Australia, including an Account of the Smelting Works in That Colony* (Adelaide: C. Platts, E. S. Wigg, G. Dehane, J. Howell, W. C. Rigby, and G. Mullett, 1863), 98.

24. Austin, *Mines of South Australia*, 99.

25. W. H. Jones, *History of the Port of Swansea*, 157.

26. Letter from "R. M." to the editor of *The Cambrian*, March 11, 1837.

27. Meeting of the Copper Companies, March 3, 1831, D/DZ207, WGAS.

28. Public Meeting of Coal Owners of Swansea and Neighbourhood, March 3, 1831, D/DZ207, WGAS.

29. Letter from J. H. Vivian to Swansea Harbour Trustees, 1839, D/DZ207, WGAS.

30. Meeting of Harbour Trustees, Swansea 1848, SW714, SCL.

31. Steven Hughes, *Copperopolis: Landscapes of the Early Industrial Period in Swansea* (Aberystwyth: Royal Commission on the Ancient and Historical Monuments of Wales, 2000), 78.

32. *The Principality*, June 21, 1850.

33. Meeting of Harbour Trustees, Swansea 1848, SW714, SCL.

34. P. W. Meik and A. O. Schenk, "Description of Swansea Harbour and of the King's Dock, Swansea," *Proceedings of the South Wales Institute of Engineers* 25, no. 1 (July 1906): 53–58.

35. Douglas Hall, *Free Jamaica 1838–1865: An Economic History* (New Haven: Yale University Press, 1959): 138–152; Roger Burt, "Virgin Gorda Copper Mine 1839–1862," *Industrial Archaeology Review* 6, no. 1 (1981): 56–62.

36. *Prospectus of The Virgin Gorda Copper Mine*, HC5.13.1, The Baring Archive.

37. John M. Smalberger, *Aspects of the History of Copper Mining in Namaqualand 1846–1931* (Cape Town: C. Struik, 1975), 127, table A; Charles K. Hyde, *Copper for America: The United States Copper Industry from Colonial Times to the 1990s* (Tucson: University of Arizona Press, 1998), 20; Otis E. Young, "Origins of the American Copper Industry," *Journal of the Early Republic* 3, no. 2 (1983): 132; Patricia Bernard Ezzell, "Burra Burra Copper Company," *The Tennessee Encyclopedia of History and Culture*, http://tennesseeencyclopedia.net/entry.php?rec=162.

38. G. T. Bloomfield, "The Kawau Copper Mine, New Zealand," *Industrial Archaeology* 11, no. 1 (1974): 1–10; Smalberger, *Aspects*, table C (folded in after 143).

39. *The Cambrian*, November 28, 1856.

40. Smalberger, *Aspects*, 105; Payton, *The Cornish Overseas*, 347–349.

41. Smalberger, *Aspects*, 69; Edmund Newell, "Taylor, John (1779–1863)," *Oxford Dictionary of National Biography* (Oxford: Oxford University Press, 2004), http://www.oxforddnd.com/view/article/27059. The Cape Copper Mining Company, which emerged after the speculative tumult of the 1850s subsided, marked a juncture between local capital, hitherto inexpertly applied, and British expertise. Behind the new firm stood John Taylor & Sons of London, the leading mining consultancy of the age.

42. Quoted in Smalberger, *Aspects*, 107.

43. Quoted in Smalberger, *Aspects*, 106.

44. Nigel Worden and Clifton Crais, "Introduction," in *Breaking the Chains: Slavery and Its Legacy in the Nineteenth-Century Cape Colony*, ed. Nigel Worden and Clifton Crais (Johannesburg: Witwatersrand University Press, 1994), 6.

45. M. K. Banton, "The Colonial Office, 1820–1855: Constantly the Subject of Small Struggles," in *Masters, Servants, and Magistrates in Britain and the Empire, 1562–1955*, ed. Douglas Hay and Paul Craven (Chapel Hill: University of North Carolina Press, 2004), 262–264.

46. "The Memorial of the Undersigned Merchants, Copper Smelters, Shipowners, Importers of Copper Ores, and Others," BPP 1847–1848 (186), 6.

47. "Smelting Copper Ore," *American Journal of Science and the Arts*, second series, 4, no. 10 (1847): 292.

48. *Exposition of the Baltimore and Cuba Smelting & Mining Company* (Baltimore: Robert Neilson, 1845). In fact, plans to exploit ores in the hinterland of Neuvitas on Cuba's north coast came to nothing. Ores from Chile and several locations in North America took their place. See also *The Mineral Industry, Its Statistics, Technology and Trade, in the United States and Other Countries to the End of 1895*, vol. 4, ed. Richard P. Rothwell (New York: The Scientific Publishing Company, 1896), 272–273.

49. Our thanks go to Ceri Carter, a doctoral candidate at the University of South Wales, who undertook genealogical research on our behalf—no easy task because most of the subjects had surnames that were exceedingly common in nineteenth-century Wales (Davies, Evans, Edwards, etc.), as they are today.

50. David Keener to R. J. Nevill, November 20, 1848, Nevill 2357, NLW, remitting money to the wives of Daniel Davies, David Bevan, Henry Charles, Evan Evans, James Evans, and Henry Roberts.

51. Collamer M. Abbott, "Isaac Tyson, Jr.: Pioneer Industrialist," *Business History Review* 42, no. 1 (1968): 67–83.

52. Young, "Origins of the American Copper industry," 117–137.

53. *Woods' Baltimore City Directory* (Baltimore: John W. Woods, 1860).

54. Benjamin S. Johns, "Henry Johns, an Early American Copper Refiner," *Engineering and Mining Journal* 93, no. 24 (June 15, 1912): 1183.

55. Hyde, *Copper for America*, 24.

56. Hyde, *Copper for America*, 24–28.

57. Roger Burt, Raymond Burnley, Michael Gill, and Alasdair Neill, *Mining in Cornwall and Devon: Mining and Men* (Exeter: Exeter University Press, 2015), 56; Newell, "The British Copper Ore Trade," 54–73.

58. Des Cowman, *The Making and Breaking of a Mining Community: The Copper Coast, County Waterford 1825–1875* (Waterford: The Mining Heritage Trust of Ireland, 2006), 128.

59. William H. Mulligan, "The Anatomy of Failure: Nineteenth-Century Irish Copper Mining in the Atlantic and Global Economy," in *The Irish in the Atlantic World*, ed. David T. Gleeson (Columbia: University of South Carolina Press, 2010), 38–52; R. A. Williams, *The Berehaven Copper Mines* (Sheffield: The Northern Mine Research Society, 1991), chapters 24–26.

60. Manuel Llorca-Jaña, "Exportaciones chilenas de cobre a Gales durante el siglo XIX: su impacto en las economías chilena y galesa," *Revista de Historia Social y de las Mentalidades* 21, no. 1 (2017): 57.

61. Luis Ortega, "Fragilities of a Frontier Zone: Townships and Villages in the Copper Mining Districts of Chile 1830–1875," unpublished paper for World of Copper workshop at Burra, South Australia, September 24–26, 2012.

62. *Adelaide Observer*, April 20, 1867.

63. *South Australian Register*, April 23, 1870.

64. Luis Valenzuela, "Challenges to the British Copper Smelting Industry in the World Market, 1840–1860," *Journal of European Economic History* 19, no. 3 (1990): 657–686, especially 676–680.

65. Valenzuela, "Challenges," 671, table 5.

66. *Morning Chronicle*, January 4, 1849.

67. *Daily News*, January 3, 1850.

68. *Daily News*, December 2, 1858.

69. *Consolidated Copper Mines of Cobre. Report of Mr. Petherick, F.G.S.* (London: George Unwin, 1863), 9–10; *The Cambrian*, October 24, 1862.

70. Jacobo de la Pezuela, *Diccionario Geografico, Estadistico, Historico de la Isla de Cuba*, 2 vols. (Madrid, 1863–1867), II, 5–14.

71. Samuel Hazard, *Cuba with Pen and Pencil* (Oxford: Signal Books, 2007), 374–375.

72. F. W. Ramsden to Commodore Phillimore, December 5, 1868, FO 72/1189, TNA.

73. *Daily News*, July 16, 1866.

74. Ada Ferrer, *Insurgent Cuba: Race, Nation, and Revolution, 1868–1898* (Chapel Hill: University of North Carolina Press, 1998); Rebecca J. Scott, *Slave Emancipation in Cuba: The Transition to Free Labor, 1860–1899* (Pittsburgh: University of Pittsburgh Press, 2000).

75. *London Gazette*, February 19, 1869.

76. *The Cambrian*, August 20, 1869.

Chapter 6 • The End of Swansea Copper, c. 1870–1924

1. E. Newell, "Copperopolis: The Rise and Fall of the Copper Industry in the Swansea District, 1826–1921," *Business History* 32, no. 3 (1990): 75.

2. E. Newell, "'The Irremediable Evil': British Copper Smelters' Collusion and the Cornish Mining Industry, 1725–1865," in *From Family Firms to Corporate Capitalism: Essays in Business and Industrial History in Honour of Peter Mathias*, ed. K. Bruland and P. O'Brien (Oxford: Oxford University Press, 1998), 173.

3. H. C. H. Carpenter, "Progress in the Metallurgy of Copper. Lecture 1," *Journal of the Royal Society of Arts* 66, no. 3398 (January 1918): 114–123.

4. Williams & Foster, Report re: possible re-location of Swansea Copper Works, September 29, 1924. LAC 126 T19, RBA.

5. See, for example, H. O'Neill, "The Invention of Bessemer in Relation to Non-ferrous Metals," *Metallurgia* (1956): 269–273.

6. A. D. Chandler, *Scale and Scope: The Dynamics of Industrial Capitalism* (Cambridge: Belknap Press of Harvard University Press, 1990), 279–281.

7. See R. R. Toomey, "Vivian and Sons, 1809–1924: A Study of the Firm in the Copper and Related Industries" (PhD thesis, University College of Swansea, 1979), 74–75.

8. For details of the Second Copper Trade Association, see Newell, "'The Irremediable Evil,'" 187–194.

9. Copy letter from H. H. Vivian to J. M. Williams, May 11, 1878, Williams & Foster Memorandum Book LAC 126 S1, RBA.

10. *The Cardiff Times*, February 15, 1868.

11. Extract from Mr. Bain's letter, Williams & Foster Memorandum Book, 42, LAC 126 S1, RBA.

12. George Grant Francis, quoted in W. H. Jones, *A History of the Port of Swansea* (Carmarthen: Spurrell, 1922), 315.

13. Andrew Ure, *A Dictionary of Arts, Manufactures and Mines; Containing a Clear Exposition of Their Principles and Practice* (New York: D. Appleton and Co., 1842), 1340; *The Cambrian*, October 25, 1834.

14. P. Claughton, "Silver and Zinc: Cardiganshire, Brittany and Dillwyn and Co. of Swansea," *Welsh Mines and Mining*, no. 2, https://www.swansea.ac.uk/media/WMM2-Claughton-Dillwyn.pdf, 45–50.

15. E. J. Cocks and B. Walters, *A History of the Zinc Smelting Industry in Britain* (London: Harrap, 1968), 14–15.

16. S. Ball, "The German Octopus: The British Metal Corporation and the Next War, 1914–1939," *Enterprise and Society* 5, no. 3 (2004); Cocks and Walters, *A History of the Zinc Smelting Industry*, 15–18.

17. Newell, "Copperopolis," 91.

18. Account of Exports and Imports in United Kingdom of Copper & c., for 12 months to 31 Dec. 1880, BPP (1881), vol. LXXXIII.

19. Toomey, "Vivian and Sons," 261; C. Evans and O. Saunders, "A World of Copper: Globalizing the Industrial Revolution, 1830–70," *Journal of Global History* 10 (2015): 20.

20. Claughton, "Silver and Zinc," 46.

21. Carpenter, "Progress in the Metallurgy of Copper: Lecture 1," 123.

22. S. J. Mackie, "On the Construction of Iron Ships and Their Preservation from Corrosion and Fouling by Zinc Sheathing," *Journal of the Society of Arts* 15, no. 753 (April 1867): 359–369.

23. A. M. Levy, "The Welsh Copper Smelting Process," *The Engineering and Mining Journal* 39 (1885): 53.

24. Toomey, "Vivian and Sons," 289.

25. *The Cambrian*, October 25, 1834.

26. R. O. Roberts, "The Smelting of Non-ferrous Metals since 1750," in *Glamorgan County History*, vol. 5, *Industrial Glamorgan*, ed. G. Williams and A. H. John (Cardiff: Glamorgan County History Trust Ltd., 1980), 88.

27. W. C. Aitken, "Brass and Brass Manufactures," in *The Industrial History of Birmingham and the Midlands Hardware District*, ed. S. Timmins (London: Robert Hardwicke, 1866), 315.

28. Return of Working Expenditure of Each Railway Company, December 1860, BPP (1861), vol. LVII.

29. Estimate from Williams and Foster, quoted in Toomey, "Vivian and Sons," 40.

30. W. P. Marshall, "Evolution of the Locomotive Engine," *Minutes of the Proceedings of the Institute of Civil Engineers* 33, no. 3 (1898): 299.

31. M. W. Kirby, "Product Proliferation in the British Locomotive Building Industry, 1850–1914: An Engineer's Paradise," *Business History* 30, no. 3 (1988): 287–305.

32. J. Morton, *Thomas Bolton and Sons Ltd., 1783–1983* (Worcester: Moorland, 1983), 35.

33. G. C. Allen, *The Industrial Development of Birmingham and the Black Country, 1860–1927* (London: Frank Cass, 1966), 51.

34. W. A. Campbell, *The Chemical Industry* (London: Longman, 1971), 17.

35. Return of Values of Imports and Exports of UK. Quantities and value of principal articles subject to duties of customs and excise, 1854–1880, BPP (1882), vol. LXVI.191.

36. S. Checkland, *The Mines of Tharsis: Roman, French and British Enterprise in Spain* (London: George Allen and Unwin, 1967), 87–96.

37. K. Davies, "William Keates of Cheadle (1801–1888) and the British Copper Industry in the Nineteenth Century," *Staffordshire Studies* 18 (2007): 37–61.

38. C. Harvey and P. Taylor, "Mineral Wealth and Economic Development: Foreign Direct Investment in Spain, 1851–1913," *Economic History Review* 40, no. 2 (1987): 189; R. Davenport-Hines, "Tennant, Sir Charles, First Baronet (1823–1906)," *Oxford Dictionary of National Biography* (Oxford: Oxford University Press, 2009), https://doi.org/10.1093/ref:odnb/36455.

39. R. B. Pettengill, "United States Foreign Trade in Copper, 1790–1932," *The American Economic Review* 25, no. 3 (September 1935): 429.

40. C. J. Schmitz, *World Non-ferrous Metal Production and Prices, 1700–1976* (London: Frank Cass, 1979), 69. Schmitz's figures, in tonnes, have been converted to tons for the purposes of this analysis.

41. K. Curtis, *Gambling on Ore: The Nature of Metal Mining in the United States, 1860–1910* (Boulder: University Press of Colorado, 2013), 117–118.

42. C. Schmitz, "The Rise of Big Business in the World Copper Industry, 1870–1930," *The Economic History Review*, new series, 39, no. 3 (1986): 392–410.

43. R. O. Roberts, "Enterprise and Capital for Non-ferrous Metal Smelting in Glamorgan, 1691–1924," *Morgannwg. Transactions of the Glamorgan Local History Society* 23 (1979): 56.

44. J. M. Smalberger, "Aspects of the History of Copper Mining in Namaqualand" (MA thesis, University of Capetown, 1969), 117.

45. Schmitz, "The Rise of Big Business," 393–396.

46. Roberts, "Enterprise and Capital," 72.

47. Ball, "The German Octopus," *Enterprise and Society* 5, no. 3 (2004): 451–489.

48. Toomey, "Vivian and Sons," 297.

49. D. M. Levy, "The Bessemerising of Copper Mattes," in *Modern Copper Smelting. Being Lectures Delivered at Birmingham University* (London: Charles Griffin and Co. Ltd., 1912), 192.

50. *The Cambrian*, November 30, 1894.

51. Newell, "Copperopolis," 93.

52. Ken Beauchamp, *History of Telegraphy* (London: The Institution of Electrical Engineers, 2001); Ben Marsden and Crosbie Smith, *Engineering Empires: A Cultural History of Technology in Nineteenth-Century Britain* (Basingstoke: Palgrave, 2007).

53. *Report of the Joint Committee Appointed by the Lords of the Committee of Privy Council for Trade and the Atlantic Telegraph Company to Enquire into the Construction of Submarine Telegraph Cables* (London: HMSO, 1861), xiii.

54. P. J. Hartog, "Matthiessen, Augustus (1831–1870), Chemist and Physicist," *Oxford Dictionary of National Biography* (Oxford: Oxford University Press, 2009), http://www.oxforddnb.com/view/10.1093/ref:odnb/9780198614128.001.0001/odnb-9780198614128-e-18349.

55. *Report of the Joint Committee*, 336.

56. Daniel R. Headrick, *The Tools of Empire: Technology and European Imperialism in the Nineteenth Century* (New York: Oxford University Press, 1981), 157–164.

57. Morton, *Thomas Bolton & Sons*, 51, 56, 59.

58. H. D. Wilkinson, *Submarine Cable Laying and Repairing* (New York: D. Van Nostrand Co., 1908), 88–89.

59. Hugh Barty-King, *Girdle Round the Earth: The Story of Cable and Wireless and Its Predecessors to Mark the Group's Jubilee 1929–1979* (London: Heinemann, 1979), 100.

60. *John A. Roebling's Sons Co., Office and Works Trenton New Jersey. Manufacturers of Iron & Steel Wire Rope and Copper Wire* [1887], 25. A copy of this catalog is at Hagley Museum & Library, Delaware.

61. Charles Bright, *Submarine Telegraphs: Their History, Construction and Working* (London: C. Lockwood and Son, 1898), 215.

62. J. Bucknall Smith, *A Treatise upon Wire, Its Manufacture and Uses* (London: Offices of *Engineering*, 1891), 102.

63. Bucknall Smith, *A Treatise upon Wire*, 100.

64. Bright, *Submarine Telegraphs*, 220.

65. Edward Dyer Peters, *The Practice of Copper Smelting* (New York: McGraw-Hill Book Co., 1911), 533.

66. For a detailed account of this, see J. Leitner, "Red Metal in the Age of Capital: The Political Ecology of Copper in the Nineteenth-Century World Economy," *Review (Fernand Braudel Centre)* 24, no. 3 (2001): 373–436.

67. D. C. Morison, "The Recent History of the Copper Trade," *Economica* 12 (November 1924): 356, 365.

68. R. Fitzgerald, "International Business and the Development of British Electrical Manufacturing, 1886–1929," *Business History Review* 91 (2017): 31–70.

69. Morison, 356.

70. J. M. Hurd, "Railways," in *The Cambridge Economic History of India*, vol. 2, *1757–1970*, ed. D. Kumar and M. Desai (Cambridge: Cambridge University Press, 1983), 749.

71. *Glasgow Herald*, December 27, 1901.

72. Aitken, "Brass and Brass Manufactures," 316.

73. K. N. Chaudhuri, "Foreign Trade and the Balance of Payments, 1757–1947," in *The Cambridge Economic History of India*, vol. 2, *1757–1970*, ed. D. Kumar and M. Desai (Cambridge: Cambridge University Press, 1983), 858.

74. *Flintshire Observer*, September 22, 1898.

75. Copy letter from John Cady (recipient unknown), July 22, 1882, Williams and Foster Memorandum Book, LAC/126/S1, RBA.

76. T. M. Ainscough, "British Trade with India," *Journal of the Royal Society of Arts* 78, no. 4044 (May 1930): 763.

77. British Copper Manufacturers, Report on the General Running of the Works, July 1926, LAC 126 W13, RBA.

78. R. Owen, *The Middle East in the World Economy, 1800–1914* (London: Methuen, 1981), 238.

79. Report by W. E. Dew on visit to German copper works, August 1924, LAC 126 T19, RBA.

80. W. H. Jones, *History of the Port of Swansea* (Carmarthen: Spurrell, 1922), 226.

81. Tables showing rates of wages paid to Morfa mill workers in 1872 and 1873, Williams and Foster Memorandum Book, 89–91, LAC 126 S1, RBA.

82. Tables showing rates of wages paid to Morfa mill workers in 1872 and 1873, Williams and Foster Memorandum Book, 89–91, LAC 126 S1, RBA.

83. R. O. Roberts, "The Development and Decline of Non-ferrous Metal Industries in South Wales," *Transactions of the Honourable Society of Cymmrodorion* (1956): 95. Although these figures are for the whole of South Wales, the vast majority were employed in the Swansea District.

84. *The Cambrian*, October 21, 1892.

85. Pascoe Grenfell and Sons Ltd., Minute Book, February 12, 1891, September 23, 1891, and October 7, 1892, LAC 45 A20, RBA.

86. E. Newell, "Grenfell Family (per. c. 1785–1879)," *Oxford Dictionary of National Biography* (Oxford: Oxford University Press, 2013).

87. *The Cambrian*, October 21, 1892.

88. Toomey, "Vivian and Sons," 247–248.

89. Letterbook of Thomas Nicholls of the Cape Copper Works, Briton Ferry, 1890–1910, LAC 72 1, RBA.

90. Letterbook of Thomas Nicholls of the Cape Copper Works, Briton Ferry, September 1890, LAC 72 1, RBA.

91. *South Wales Daily News* report, quoted in L. J. Williams, "The New Unionism in South Wales, 1882–92," *Welsh History Review* 1, no. 4 (1963): 413.

92. Williams and Foster Memorandum Book, copy letter to Mr. Martin, May 13, 1873, LAC 126 S1, RBA.

93. P. S. Thomas, "Industrial Relations," *Social and Economic Survey of Swansea and District*, pamphlet no. 3 (Swansea, 1940), 42.

94. *Y Gwladgarwr*, June 14, 1873.

95. D. Minister, "Trade Union Activity in the Tinplate, Nickel and Coal Industries in the Swansea Valley, c. 1870–1926" (unpublished PhD thesis, Swansea University, 2009), 85–87.

96. *The Cambrian*, July 11, 1890.

97. Letterbook of Thomas Nicholls, September 1890, LAC 72 1, RBA.

98. *The Cambrian*, September 23, 1904.

99. *The Cambrian*, April 21, 1905.

100. See, for example, *Evening Express*, February 16, 1903; *Cardiff Times*, December 26, 1903; *Cambria Daily Leader*, June 2, 1911.

101. *The Cambrian*, August 1, 1890.

102. *Cambria Daily Leader*, January 13, 1915.

103. *Cambria Daily Leader*, May 12, 1919.

104. S. Broadberry and P. Howlett, "The United Kingdom during World War 1: Business as Usual?," in S. Broadberry and M. Harrison, eds., *The Economics of World War 1* (Cambridge: Cambridge University Press, 2005), 212.

105. N. Delaney, "The Great War and the Transformation of the Atlantic Copper Trade," *Scandinavian Economic History Review* 65, no. 3 (2017): 263–278 (271).

106. *Cambria Daily Leader*, April 4, 1917.

107. Delaney, "The Great War."

108. Final Report of the Committee on Commercial and Industrial Policy after the War, 15–16, BPP (1918), vol. XIII, 239.

109. Ball, "The German Octopus," 461.

110. Imperial War Conference, 1918. Extracts from minutes of proceedings and papers, Sir Albert Stanley, BPP (1918).

111. Morison, "The Recent History of the Copper Trade," 357–361.

112. Smalberger, "Aspects," 137.

113. For details, see Morison, "The Recent History of the Copper Trade," 361–362.

114. Minutes of Meetings of Vivian and Sons and British Copper Manufacturers, Report on the accounts of Williams and Foster for 9-1/2 months ending mid-May 1924 and on the proposed amalgamation with Vivian and Sons Ltd., LAC 126 K3, RBA.

115. Minutes of Meetings of Vivian and Sons and British Copper Manufacturers, Report on the accounts of Williams and Foster, LAC 126 K3, RBA.

116. Minutes of Meetings of Vivian and Sons and British Copper Manufacturers, Meeting of Directors of Vivian and Sons Ltd., London, July 10, 1924, LAC 126 K3, RBA.

117. Minutes of Meetings of Vivian and Sons and British Copper Manufacturers, Meeting of Directors of Vivian and Sons Ltd., London, July 10, 1924, LAC 126 K3, RBA.

118. Imperial Chemical Industries Ltd., Metals Division, Landore Swansea 1948, Details of Copper Specifications on Hand, LAC 126 V6, RBA.

119. W. J. Reader, *Imperial Chemical Industries. A History*, vol. 2, *The First Quarter Century, 1926–1952* (London, 1975), 13–14, 468.

120. J. Barr, *Derelict Britain* (Harmondsworth: Penguin, 1969), 79.

121. M. Chisholm and J. Howells, "Derelict Land in Great Britain. A Context for the Lower Swansea Valley," in R. Bromley and G. Humphreys, eds., *Dealing with*

Dereliction. The Re-development of the Lower Swansea Valley (Swansea: University College of Swansea, 1979), 10.

122. K. J. Hilton, ed., *The Lower Swansea Valley Project* (London: Longmans, 1967).

123. B. J. Morin, *The Legacy of American Copper Smelting: Industrial Heritage Versus Environmental Policy* (Knoxville: University of Tennessee Press, 2013), 120.

124. Barr, *Derelict Britain*, 104–105.

125. John Young, "Swansea Gives New Life to Its Blighted Valley: Wasteland Devastated by Copper Fumes Reclaimed in Pace-Setting Project," *The Times*, February 11, 1981.

126. Examples include Rhondda Heritage Park, opened on the site of the former Lewis Merthyr Colliery in Trehafod, in 1989; and Big Pit, developed as a coal mining museum in Blaenavon in 1983 but acquired by the National Museum of Wales in 1999.

127. C. S. Briggs, "The Future of Industrial Archaeology in Wales," in C. S. Briggs, ed., *Welsh Industrial Heritage: A Review*, CBA Research Report no. 79 (London: Council for British Archaeology, 1992), 145.

128. H. V. Bowen, "Copperopolis: Swansea's Heyday, Decline and Regeneration," Legatum Institute Lecture, March 16, 2016, 10.

Chapter 7 • Swansea Copper in World-Historical Perspective

1. David Killick and Thomas Fenn, "Archaeometallurgy: The Study of Preindustrial Mining and Metallurgy," *Annual Review of Anthropology* 41 (2012): 563.

2. The reduction of copper ores in sub-Saharan Africa comes substantially later, in the first millennium BCE: David Killick, "A Global Perspective on the Pyrotechnologies of Sub-Saharan Africa," *Azania: Archaeological Research in Africa* 51, no. 1 (2016): 62–87. In the western hemisphere, the smelting of copper began at roughly the same time: A. Eichler, G. Gramich, T. Kellerhals, L. Tobler, Th. Rehren, and M. Schwikowski, "Ice-Core Evidence of Earliest Extensive Copper Metallurgy in the Andes 2700 Years Ago," *Scientific Reports* 7, article no. 41855; doi:10.1038/srep41855, January 31, 2017.

3. *The City of Swansea: Challenges and* Change, ed. Ralph A. Griffiths (Gloucester: Sutton Publishing, 1990), 40; *Swansea: An Illustrated History*, ed. Glanmor Williams (Swansea: C. Davies, 1990), 32.

4. Sungmin Hong, Jean-Pierre Candelone, Michel Soutif, and Claude F. Boutron, "A Reconstruction of Changes in Copper Production and Copper Emissions to the Atmosphere during the Past 7000 Years," *The Science of the Total Environment* 188 (1996): 183–193; Sungmin Hong, Jean-Pierre Candelone, Clair C. Patterson, and Claude F. Boutron, "History of Ancient Copper Smelting Pollution during Roman and Medieval Times Recorded in Greenland Ice," *Science* 272 (April 12, 1996): 246–249.

5. Robin Fleming, "Recycling in Britain after the Fall of Rome's Metal Economy," *Past and Present* 217 (2012): 3–45.

6. Ekkehard Westermann, "Tendencies in the European Copper Market in the 15th and 16th Centuries," in *Precious Metals in the Age of Expansion: Papers of the XIVth International Congress of the Historical Sciences*, ed. Hermann Kellenbenz (Stuttgart: Klett-Cotta, 1981), 71–77.

7. Ragnhild Hutchison, "Exploring an Early Cross-Border Trade System: Norwegian Copper in the 18th Century," *Scandinavian Journal of History* (October 6,

2018); Kristin Ranestad, "Copper Trade and Production of Copper, Brass and Bronze Goods in the Oldenburg Monarchy: Copperworks and Copper Users in the Eighteenth Century," *Scandinavian Economic History Review* (January 13, 2019).

8. Thomas R. Fenn, David J. Killick, John Chesley, Sonja Magnavita, and Joaquin Ruiz, "Contacts between West Africa and Roman North Africa: Archaeometallurgical Results from Kissi, Northeastern Burkina Faso," in *Crossroads/Carrefour Sahel: Cultural and Technological Developments in First Millennium BC/AD West Africa*, ed. Sonja Magnavita, Lassina Koté, Peter Breunig, and Oumarou A. Idé (Frankfurt: Africa Magna Verlag, 2009), 119–146; Victoria Leitch, Chloë Duckworth, Aurélie Cuénod, David Mattingly, Martin Sterry, and Franca Cole, "Early Saharan Trade: The Inorganic Evidence," in *Trade in the Ancient Sahara and Beyond*, ed. D. J. Mattingly, V. Leitch, C. N. Duckworth, A. Cuénod, M. Sterry, and F. Cole (Cambridge: Cambridge University Press, 2017), 287–340, and Sonja Magnavita, "Track and Trace: Archaeometric Approaches to the Study of Early Trans-Saharan Trade," in the same volume, 393–413.

9. Sarah M. Guérin, "Gold, Ivory, and Copper: Materials and Arts of Trans-Saharan Trade," in *Caravans of Gold, Fragments in Time: Art, Culture, and Exchange across Medieval Saharan Africa*, ed. Kathleen Bickford Berzock (Princeton: Block Museum of Art, Northwestern University, in association with Princeton University Press, 2019), 175–201; Raymond Silverman, "Red Gold: Things Made of Copper, Bronze, and Brass," in the same volume, 257–267.

10. Théodore Monod, "Le Maden Ijâfen: Une épave caravanière ancienne dans la Majâbat al-Koubrâ," in *Actes du premier colloque international d'archéologie africaine, Fort- Lamy (République du Tchad) 11–16 décembre 1966* (Fort-Lamy: Institut national tchadien pour les sciences humaines, 1969), 286–320.

11. Eugenia W. Herbert, *Red Gold of Africa: Copper in Precolonial History and Culture* (Madison: University of Wisconsin Press), 127.

12. Shadreck Chirikure, Ashton Sinamai, Esther Goagose, Marina Mubusisi, and W. Ndoro, "Maritime Archaeology and Trans-Oceanic Trade: A Case Study of the Oranjemund Shipwreck Cargo, Namibia," *Journal of Marine Archaeology* 5 (2010): 37–55.

13. K. N. Chaudhuri, *The Trading World of Asia and the English East India Company 1660–1760* (Cambridge: Cambridge University Press, 1978), 206, 221.

14. Ryuto Shimada, *The Intra-Asian Trade in Japanese Copper by the Dutch East India Company during the Eighteenth Century* (Leiden: Brill, 2006), 172.

15. Stephen N. Broadberry, Bruce M. S. Campbell, and Bas van Leeuwen, "When Did Britain Industrialise? The Sectoral Distribution of the Labour Force and Labour Productivity in Britain, 1381–1851," *Explorations in Economic History* 50, no. 1 (2013): 16–27.

16. Jules Ginswick, ed., *Labour and the Poor in England and Wales, 1849–1851: The Letters to* The Morning Chronicle *from the Correspondents in the Manufacturing and Mining Districts, the Towns of Liverpool and Birmingham, and the Rural Districts*, vol. 3, *The Mining and Manufacturing Districts of South Wales and North Wales* (London: Frank Cass & Co., 1983), 184.

17. The high-wage thesis is presented in convenient form in Robert C. Allen, *The British Industrial Revolution in Global Perspective* (Cambridge: Cambridge University Press, 2009).

18. The literature is extensive, but for two of the most effective interventions, see John Styles, "Fashion, Textiles and the Origins of the Industrial Revolution," *East Asian Journal of British History* 5 (2016): 161–190; and Jane Humphries and Benjamin Schneider, "Spinning the Industrial Revolution," *Economic History Review* 72, no. 1 (2019): 126–155.

19. John Morton, "The Rise of the Modern Copper and Brass Industry in Britain 1690–1750" (PhD thesis, University of Birmingham, 1985), appendix B (5).

20. Jin Cao, "The Last Copper Century: Southwest China and the Coin Economy (1705–1808)," *Asian Review of World Histories* 7, nos. 1–2 (2019): 126–146.

21. Miroslav Lacko, "Copper Production in the Habsburg Monarchy during the 18th Century—Quantification of the Production and Profitability," paper at the Fifth ENIUGH Congress, Budapest, 2017.

22. A. Snowden Piggot, *The Chemistry and Metallurgy of Copper, including a Description of the Principal Copper Mines of the United States and other Countries* (Philadelphia: Lindsay & Blakiston, 1858), 202.

23. Alf Zachäus, *Mansfeld and the German Economy in the Nineteenth Century* (Munich: GRIN Verlag, 2015).

24. John Scoffern, William Truran, William Clay, Robert Oxland, William Fairbairn, W. C. Aitkin, and William Vose Pickett, *The Useful Metals and Their Alloys, including Mining Ventilation, Mining Jurisprudence, and Metallic Chemistry Employed in the Conversion of Iron, Copper, Tin, Zinc, Antimony and Lead Ores; with Their Application to the Industrial Arts* (London: Houlston & Wright, 1866), 560.

25. We borrow the phrase "actually existing capitalism" from the editors' introduction to *American Capitalism: New Histories*, ed. Sven Beckert and Christine Desan (New York: Columbia University Press, 2018).

26. Philip Jenkins, *The Making of a Ruling Class: The Glamorgan Gentry 1640–1790* (Cambridge: Cambridge University Press, 1983), 16–17, 47–60.

27. A. H. John, "War and the English Economy, 1700–1763," *Economic History Review* 7, no. 3 (1955): 329–344.

28. Morton, "The Rise of the Modern Copper and Brass Industry," 190.

29. Chris Evans, *Slave Wales: The Welsh and Atlantic Slavery, 1660–1850* (Cardiff: University of Wales Press, 2010), 31–34.

30. Edward Dyer Peters, *The Practice of Copper Smelting* (New York: McGraw-Hill Book Co., 1911), 533.

31. Timothy J. LeCain, *Mass Destruction: The Men and Giant Mines That Wired America and Scarred the Planet* (New Brunswick: Rutgers University Press, 2009), 129–137.

32. Peter Bell and Justin McCarthy, "The Evolution of Early Copper Smelting Technology in Australia (Part II)," *Journal of Australasian Mining History* 9 (2011): 31.

INDEX

Page numbers in italics indicate figures and tables.